Geometry for Prin

Geometry for Primary Grade 1

 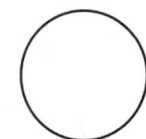

Helping students form an understanding of geometric shapes is a challenging task. In order to help students learn to recognize shapes, they must be approached in terms that will have meaning for them. The National Council of Teachers of Mathematics (NCTM) has set specific standards to help students become confident in their mathematical abilities. Geometry is an important component of the primary mathematics curriculum because geometric knowledge, relationships, and insights are useful in everyday situations and are connected to other mathematical topics and school subjects. *Geometry for Primary Grade 1* blends the vision of the NCTM Standards.

Geometry for Primary Grade 1 provides you with the opportunity to expand students' knowledge of geometry. Activities requiring the identification of sides, corners, line segments, and congruent figures are presented. Students discriminate between open and closed figures, and recognize the shapes and names for solid and plane figures. They also identify the number of sides and corners of squares, rectangles, and triangles. Identifying same size and same shape and recognizing lines of symmetry are skills included in this book. Students practice completing series of geometric patterns. They continue their study by measuring with centimeters and inches, finding perimeter and area, and learning to identify fractions and fractional parts of a whole.

Art activities that involve symmetry will enhance the learning experience for most students, as will the discussion of shapes in the classroom or a scavenger hunt for shapes throughout the classroom. Providing experiences for exploration of open and closed or symmetrical alphabet letters provides cross-curricular fun.

It is essential that students be given sufficient concrete examples of geometric concepts. Manipulatives that can be used to reinforce the skills are recommended on the activity pages.

Organization

Seven units cover the basic geometric skills presented in the first grade: basic ideas of geometry; plane figures; solid figures; symmetry; measurement; patterns; and fractions using pictorial models. In *Geometry for Primary Grade 1*, the mathematics curriculum is presented so that students can:

- formulate and solve problems from everyday and mathematical situations

- describe, draw, and classify shapes

- investigate and predict the results of combining, subdividing, and changing shapes

- develop spatial sense

- relate geometric ideas to number and measurement ideas

- recognize and appreciate geometry in their world

- understand the attributes of length, perimeter, and area

- develop concepts of fractions and mixed numbers

- recognize, describe, and extend a variety of patterns.

Geometry for Primary Grade 1

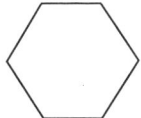  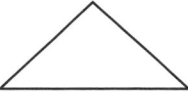 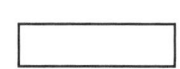 

Use

The activities in this book are designed for independent use by students who have had instruction in the specific skills covered in the lessons. Copies of the activity sheets can be given to individuals or pairs of students for completion. When students are familiar with the content of the worksheets, they can be assigned as homework.

To begin, determine the implementation that fits your students' needs and your classroom structure. The following plan suggests a format for this implementation.

1. **Administer** the Assessment tests to establish baseline information on each student. These tests may also be used as post-tests when students have completed a unit.

2. **Explain** the purpose of the worksheets to the class.

3. **Review** the mechanics of how you want students to work with the activities. Do you want them to work in pairs? Are the activities for homework?

4. **Introduce** students to the process and purpose of the activities. Work with students when they have difficulty. Give them only a few pages at a time to avoid pressure.

Additional Notes

1. <u>Parent Communication</u>. Send the Letter to Parents home with students.

2. <u>Student Communication</u>. Encourage students to share the Letter to Students with their parents.

3. <u>Manipulatives</u>. Manipulatives are recommended at the bottom of the activity pages. This optional feature can help you provide concrete examples to reinforce geometric concepts.

4. <u>NCTM Standards Correlation</u>. This chart helps you with your lesson planning. An icon for each standard is included on the student page so that you can tell at a glance what skill is being reinforced on the page.

5. <u>Student Progress Chart</u>. Duplicate the grid sheets found on pages 6-7. Record student names in the left column. Note date of completion of each lesson for each student.

6. <u>Have fun</u>! Working with these activities can be fun as well as meaningful for you and your students.

Dear Parent:

During this school year, our class will be working with mathematical skills. We will be completing activity sheets that provide enrichment in the area of geometry. This includes skills in problem solving, geometry and spatial sense, fractions, and patterns.

From time to time, I may send home activity sheets. To best help your child, please consider the following suggestions:

- Provide a quiet place to work.
- Go over the directions together.
- Encourage your child to do his or her best.
- Check the lesson when it is complete.
- Go over your child's work, and note improvements as well as problems.

Help your child maintain a positive attitude about mathematics. Let your child know that each lesson provides an opportunity to have fun and to learn. If your child expresses anxiety about these strategies, help him or her understand what causes the stress. Then talk about ways to eliminate math anxiety.

Above all, enjoy this time you spend with your child. He or she will feel your support, and skills will improve with each activity completed.

Thank you for your help!

Cordially,

Dear Student:

This year you will be working in many areas in mathematics. The activities in this program concentrate on the area of geometry. You will work with lines, sides, and corners; plane and solid figures; matching shapes that are the same shape and size; lines of symmetry; measuring the distance around and the area of figures; equal parts of a figure; patterns; and using a graph to find spots on a map. You will get to color, draw, count, and sort shapes, measure size, draw patterns, and solve problems. These activities will show you fun ways to practice geometry!

As you complete the worksheets, remember the following:

- Read the directions carefully.
- Read each question carefully.
- Check your answers after you complete the activity.

You will learn many ways to solve math problems. Have fun as you develop these skills!

Sincerely,

STUDENT PROGRESS CHART

STUDENT NAME	UNIT 1 BASIC IDEAS OF GEOMETRY								UNIT 2 PLANE FIGURES																	UNIT 3 SOLID FIGURES											UNIT 4 SYMMETRY					
	10	11	12	13	14	15	16	18	19	20	21	22	23	24	25	26	27	28	29	30	31	32	33	35	36	37	38	39	40	41	42	43	44	45	47	48	49	50	51

STUDENT PROGRESS CHART

STUDENT NAME	UNIT 5 MEASUREMENT														UNIT 6 PATTERNS							UNIT 7 FRACTIONS USING PICTORIAL MODELS																					
	53	54	55	56	57	58	59	60	61	62	63	64	65	66	67	69	70	71	72	73	75	76	77	78	79	80	81	82	83	84	85	86	87	88	89	90	91	92	93	94

NCTM STANDARDS CORRELATION

NCTM Standard	Unit 1	Unit 2	Unit 3	Unit 4	Unit 5	Unit 6	Unit 7
1: Problem Solving • formulate problems from everyday and mathematical situations	16	30, 31, 32, 33	44, 45		67		79, 92, 93, 94
9: Geometry & Spatial Sense • describe shapes	16		35, 36				
9: Geometry & Spatial Sense • draw shapes		26, 27, 28, 30, 33			62, 67		94
9: Geometry & Spatial Sense • classify shapes		18, 19, 20, 21, 22, 23, 24, 25, 29, 32	37, 38, 39, 40, 41, 42, 43, 44, 45				
9: Geometry & Spatial Sense • combine, subdivide, and change shapes		33		47, 48, 49, 50, 51			75, 76, 77, 78, 79, 80, 81, 82, 83, 84, 85, 86
9: Geometry & Spatial Sense • develop spatial sense	10, 11, 12, 13, 14, 15, 16	19, 21, 22, 23, 24, 25, 26, 27, 32	35, 36, 44	47, 48, 49, 50, 51		69, 70, 71, 72, 73	75, 76, 77, 78, 79
9: Geometry & Spatial Sense • relate geometric ideas to number and measurement ideas	13, 14, 15, 16	28, 29, 32	38, 43		62, 63, 64, 65, 66, 67	71	75, 76, 77, 78, 79, 80, 81, 82, 83, 84, 85, 86, 87, 88, 89, 90, 91, 92, 93, 94
9: Geometry & Spatial Sense • recognize and appreciate geometry in their world		19, 32	37, 39, 40, 41, 42, 43	49, 51			79, 92, 93, 94
10: Measurement • understand the attributes of length					53, 54, 55, 56, 57, 58, 59, 60, 61, 62, 63, 64, 66		
10: Measurement • understand the attributes of area					65, 66, 67		
12: Fractions & Decimals • develop concepts of fractions and mixed numbers							75, 76, 77, 78, 79, 80, 81, 82, 83, 84, 85, 86, 87, 88, 89, 90, 91, 92, 93, 94
13: Patterns & Relationships • recognize, describe, and extend a variety of patterns						69, 70, 71, 72, 73	

Name _____ Date _____

BASIC IDEAS OF GEOMETRY

Assessment: Unit 1

Color inside each closed figure.
Cross out the figures that are open.

1.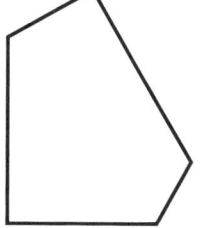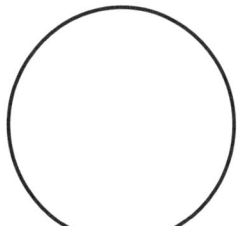

Trace each side blue.

Draw a ⬭ on each corner.

Write how many sides and corners.

2. 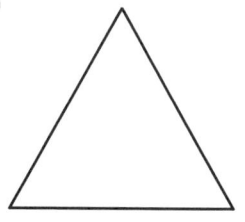 ____ sides

____ corners

3. 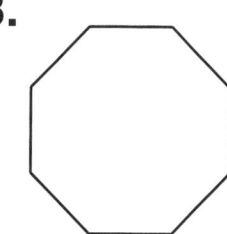 ____ sides

____ corners

circle   rectangle   triangle   square

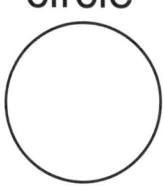 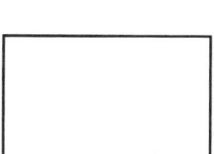  

4. I have 3 sides.
I have 3 corners.

I am a

_____

- - - - - - - - - - - - - - - - - - - .

5. I have no sides.
I have no corners.

I am a

_____

- - - - - - - - - - - - - - - - - - - .

Name _____ Date _____

Open and Closed Figures

Color inside each closed figure.
Ring the figures that are open.

 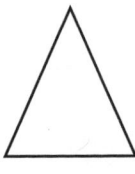

open   closed

1.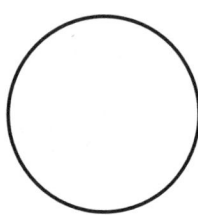

2.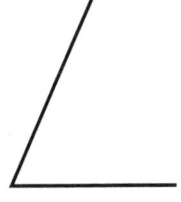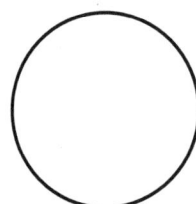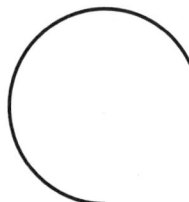

Color inside each rectangle.

3.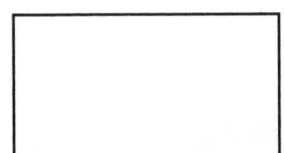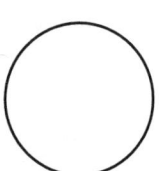

Use Pipe Cleaners to make open and closed figures. ▮

Name _____ Date _____

Open and Closed Figures

1. Draw each shape on the correct tree.

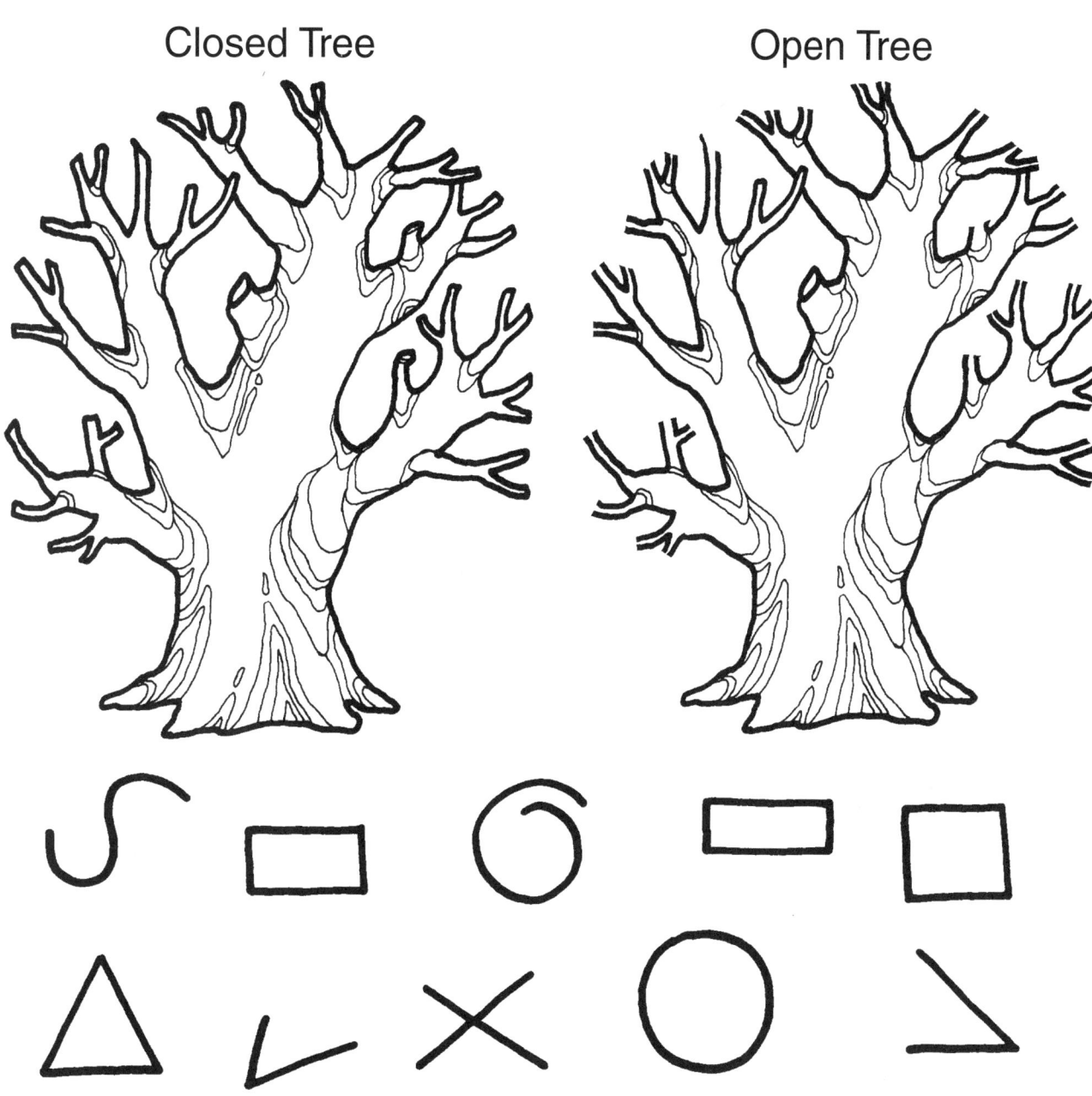

Closed Tree   Open Tree

2. Color inside each closed shape.

Sort Pipe Cleaners into open and closed figures.

BASIC IDEAS OF GEOMETRY

Sides and Corners

Trace each side blue.

Draw a ◯ on each corner.

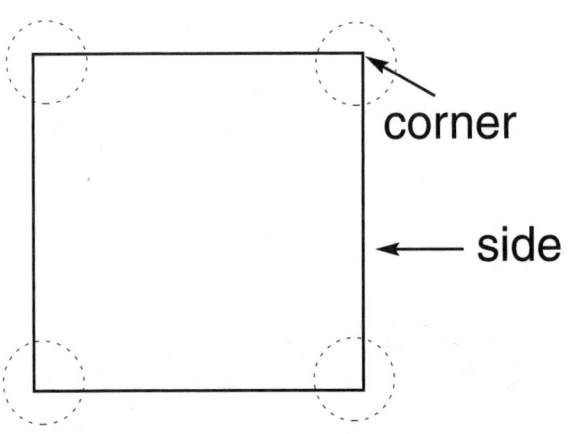

corner

side

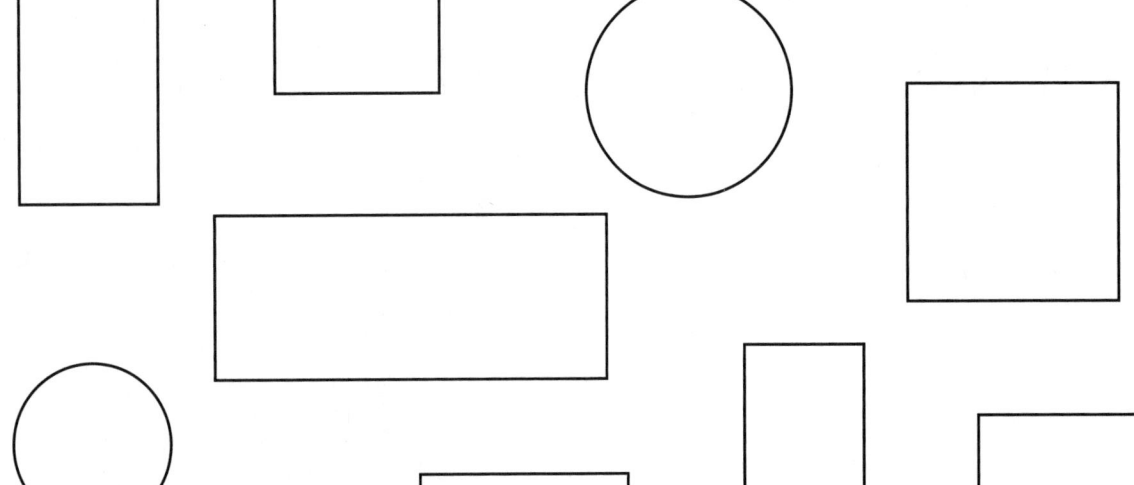

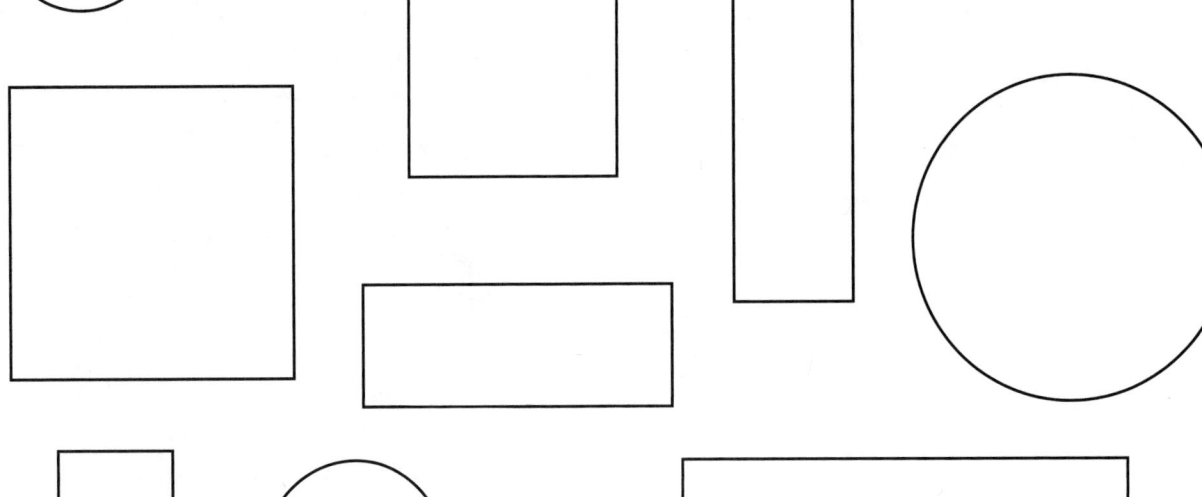

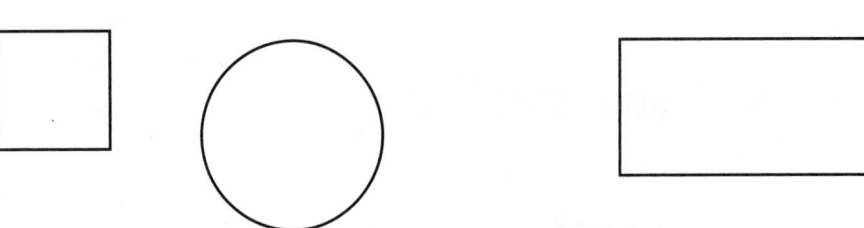

Match Attribute Blocks to shapes on page.

Name _____ Date _____

Sides and Corners

Trace each side blue.

Draw a ⬭ on each corner.

Write how many sides and corners.

1. corner
side
_4_ sides
_4_ corners

2.
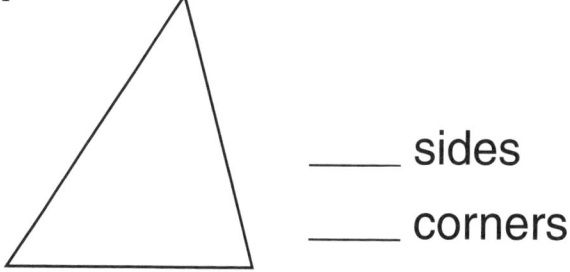
___ sides
___ corners

3.
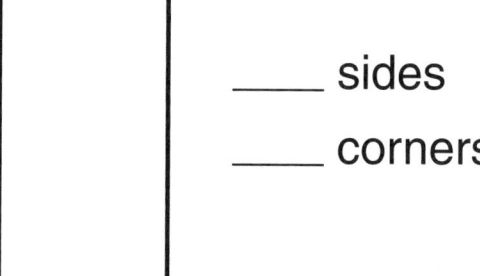
___ sides
___ corners

4.
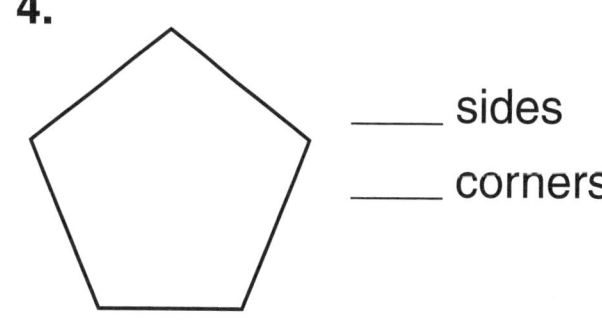
___ sides
___ corners

5.
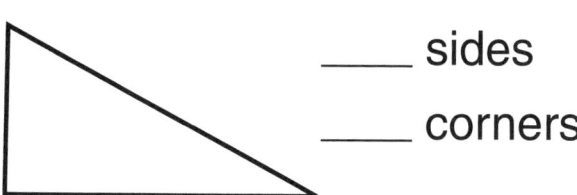
___ sides
___ corners

6.
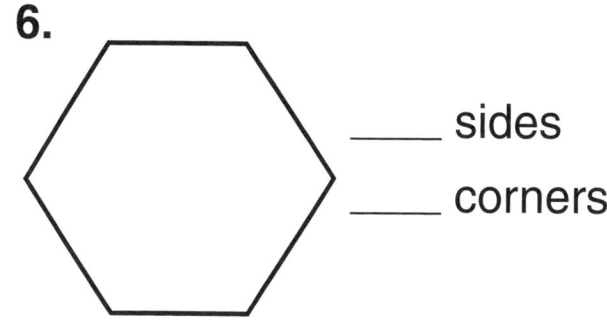
___ sides
___ corners

Sort Attribute Blocks by number of sides and corners. ⬭

Name _____ Date _____

Sides and Corners

Trace each side ⬛ blue. ▷

Draw a ⬛ red ▷ ◯ on each corner.

Write how many sides and corners.

1.

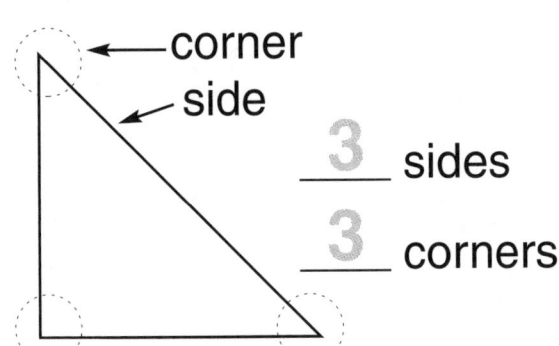

_3_ sides

_3_ corners

2.

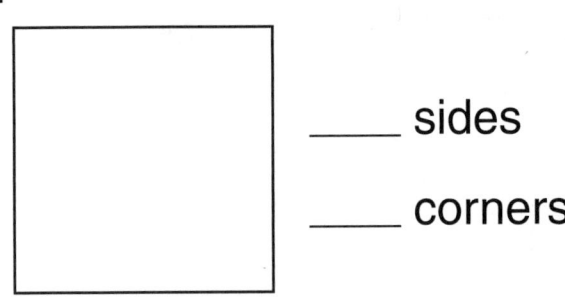

____ sides

____ corners

3.

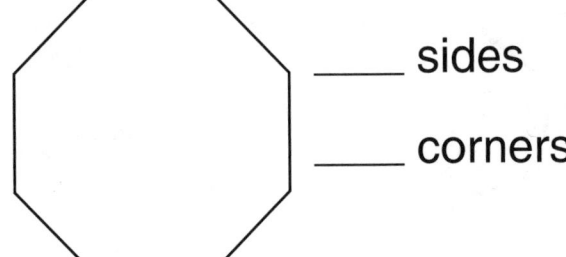

____ sides

____ corners

4.

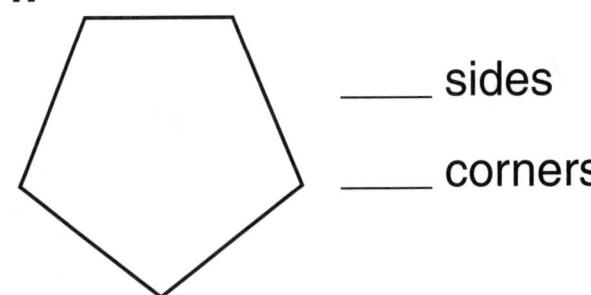

____ sides

____ corners

5.

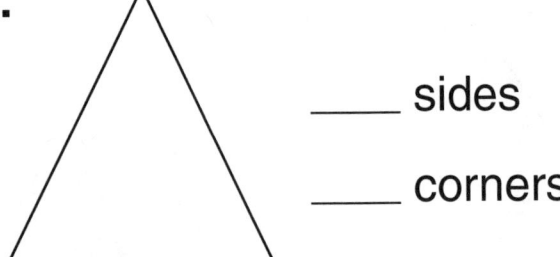

____ sides

____ corners

6.

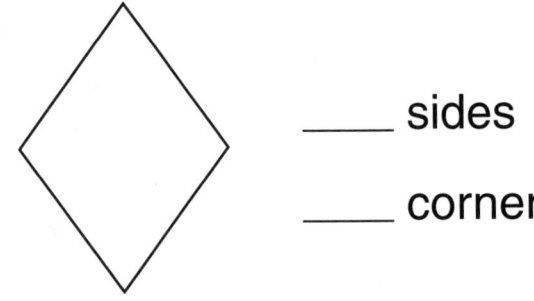

____ sides

____ corners

Sort Attribute Blocks by number of sides and corners. ⬭

BASIC IDEAS OF GEOMETRY

Sides, Corners, and Square Corners

A rectangle has 4 sides and 4 square corners.

....

Write how many sides, corners, and square corners.

1.

___4___ sides

___4___ corners

___4___ square corners

2.

_____ sides

_____ corners

_____ square corners

3.

_____ sides

_____ corners

_____ square corners

4.

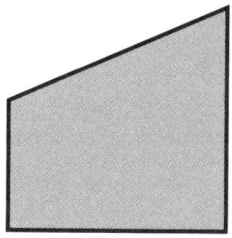

_____ sides

_____ corners

_____ square corners

5.

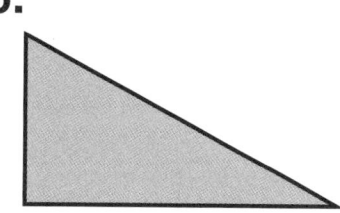

_____ sides

_____ corners

_____ square corners

6.

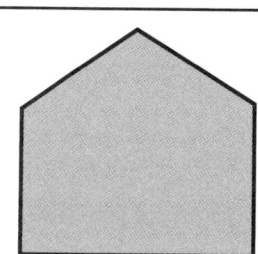

_____ sides

_____ corners

_____ square corners

Match Attribute Blocks to shapes on page.

BASIC IDEAS OF GEOMETRY

Problem Solving

square    circle 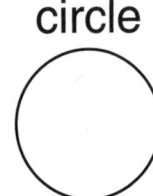   triangle 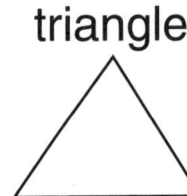   rectangle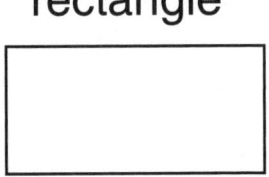

....

Listen carefully. Write my name.

1. I have 4 sides.
I have 4 corners.
All 4 sides are the
same length.

I am a

_____

- - - - - - - - - - - - - -

_____.

2. I have 4 sides.
I have 4 corners.
I have 2 long sides.
I have 2 short sides.

I am a

_____

- - - - - - - - - - - - - -

_____.

3. I have 3 sides.
I have 3 corners.

I am a

_____

- - - - - - - - - - - - - -

_____.

4. I have no sides.
I have no corners.

I am a

_____

- - - - - - - - - - - - - -

_____.

Duplicate shapes on Geoboards. ❓ ▢

Name _____   Date _____

Assessment: Unit 2

Ring the same shape.

1.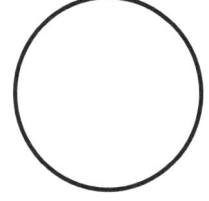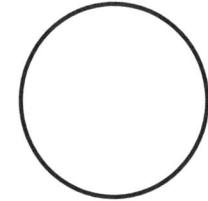

2.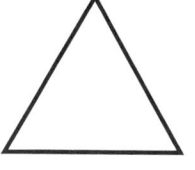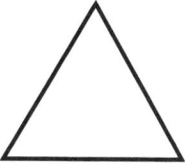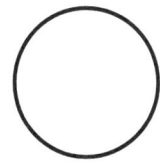

3.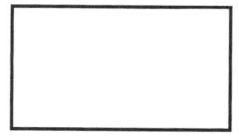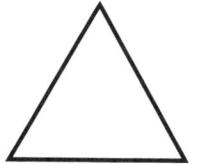

4.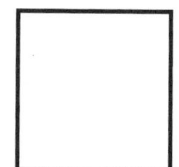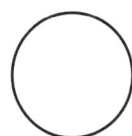

Ring the one that is the same size and shape.

5.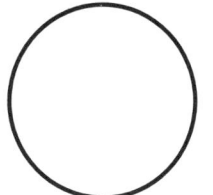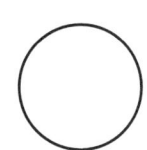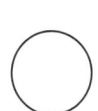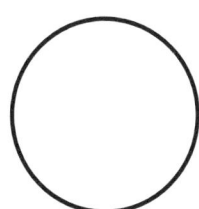

PLANE FIGURES

Identifying Same Shape

triangle

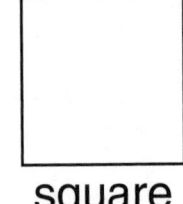
square

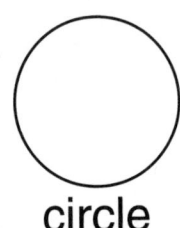
circle

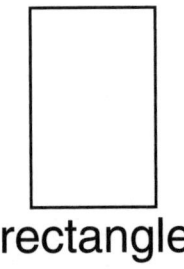
rectangle

Ring the same shape.

1.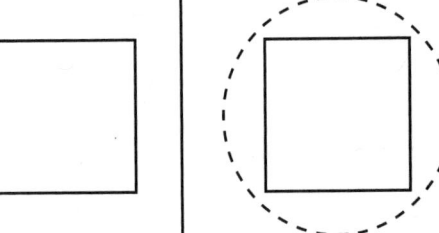

2.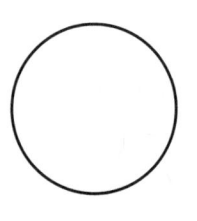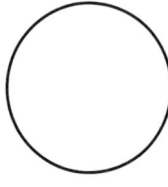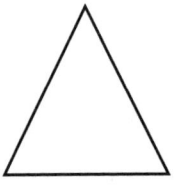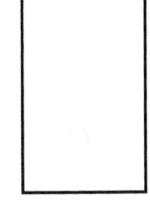

3.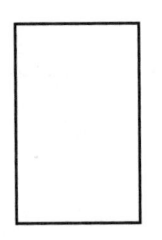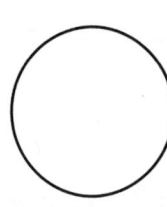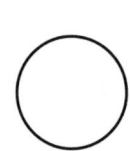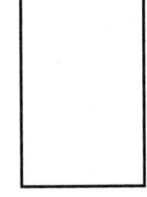

4.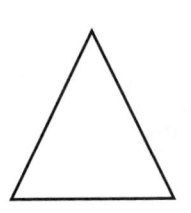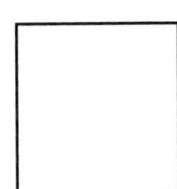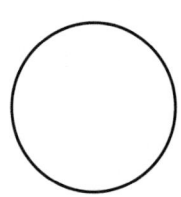

Sort Attribute Blocks by shape.

PLANE FIGURES

Identifying Plane Figures

 Color triangles RED.

Color rectangles GREEN.

Color circles BLUE.

Color squares YELLOW.

Make pictures using Tangram Puzzles.

Name _____ Date _____

Matching Plane Figures

Draw lines to match the shapes that are the same.

1.

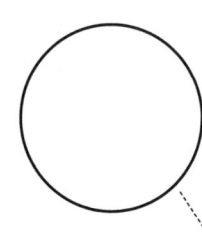  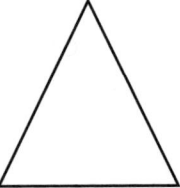

2.

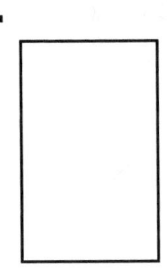  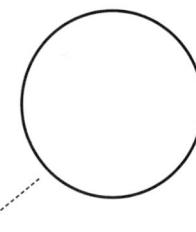

3.

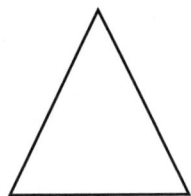 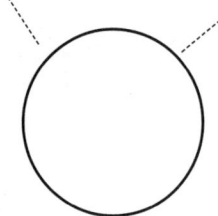 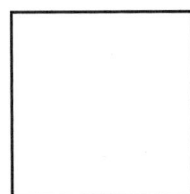

4.

  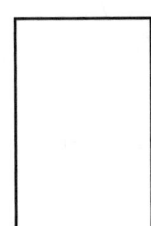

Sort Attribute Blocks by shape. ⬛

Name _____   Date _____

PLANE FIGURES

Matching Size and Shape

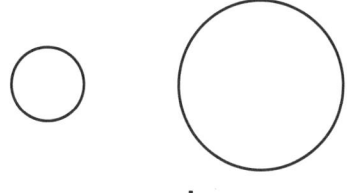

same shape

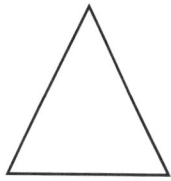 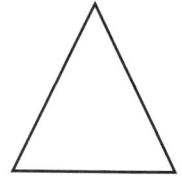

same size and same shape

Color the one that is the same size and
the same shape.

1.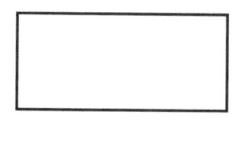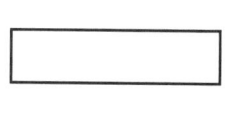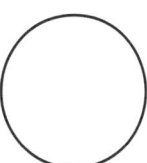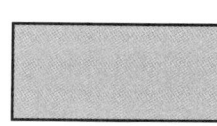

2.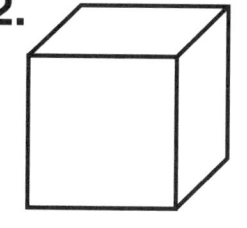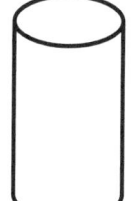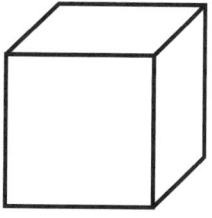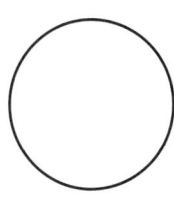

3.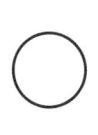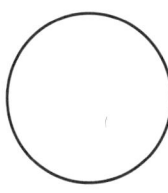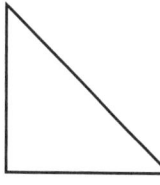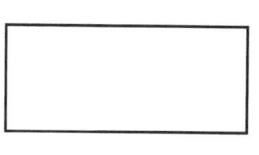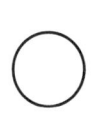

4.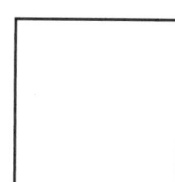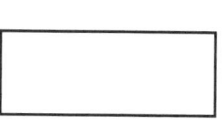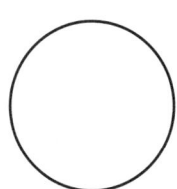

Sort Attribute Blocks by shape. ⬭

PLANE FIGURES

Matching Size and Shape

Color the same size and shape.

1.

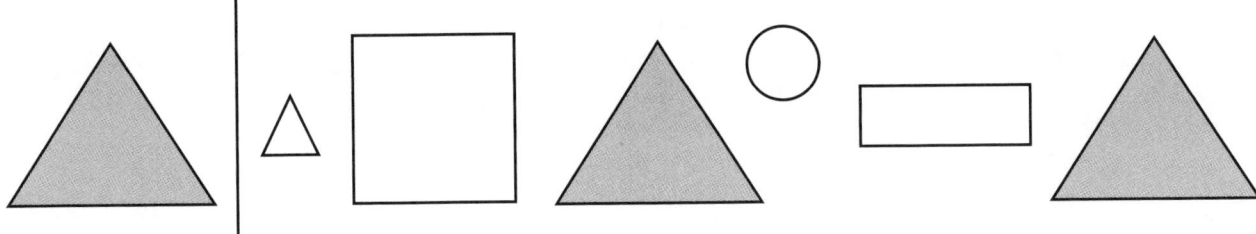

2.

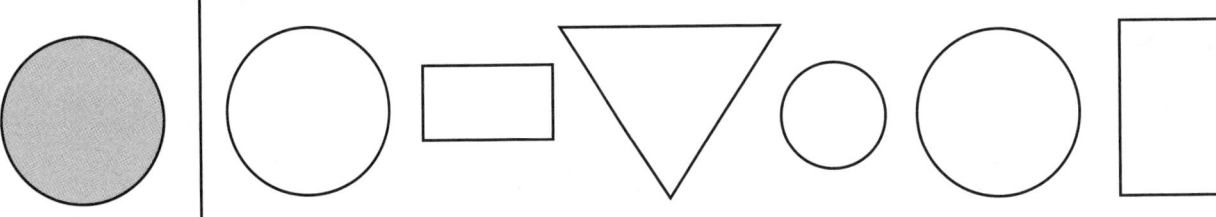

3.

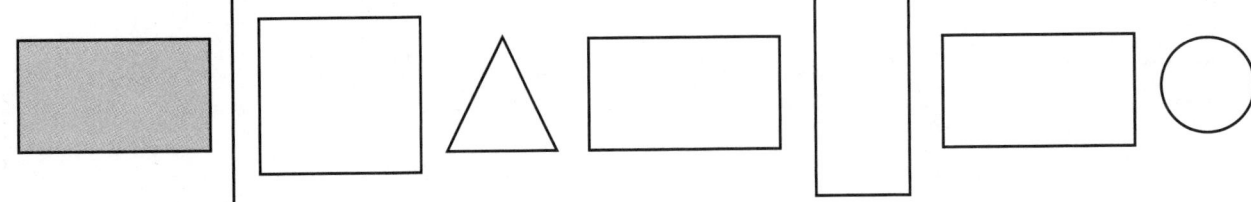

4.

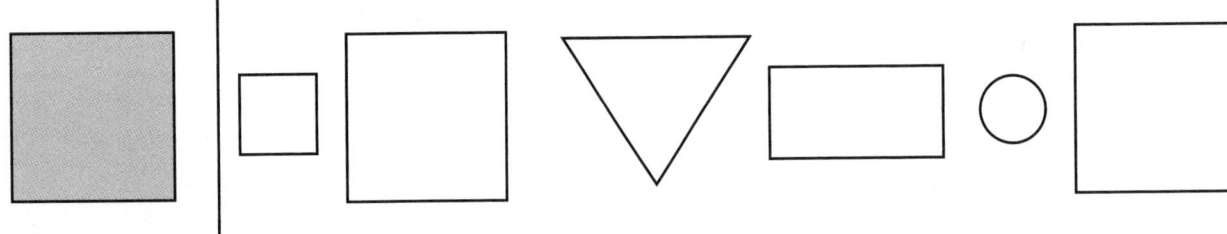

Sort Attribute Blocks by shape. ⬛

PLANE FIGURES

Matching Size and Shape

Ring the one that is the same size and shape.

1.

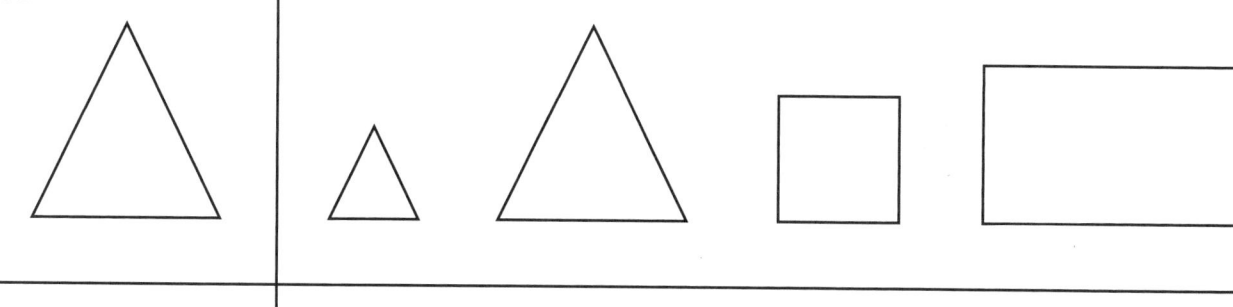

2.

3.

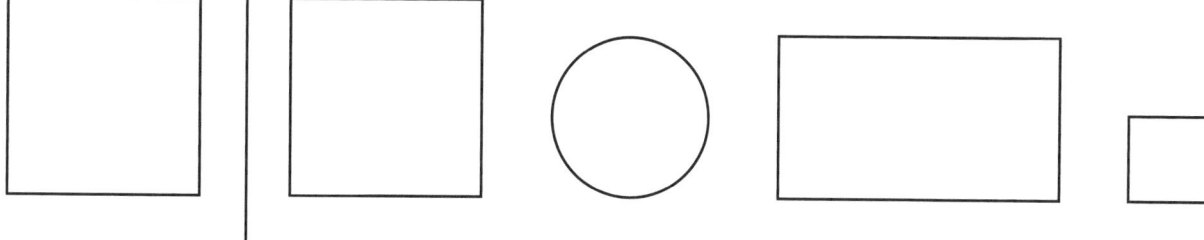

4.

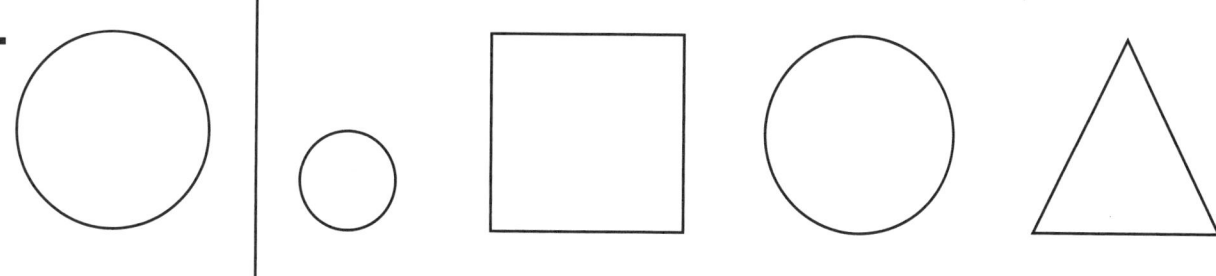

Sort Attribute Blocks by shape. ⬭

Name _____   Date _____

Matching Size and Shape

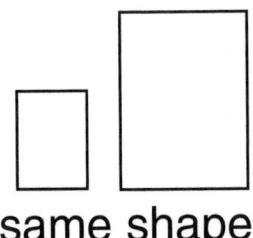

same shape

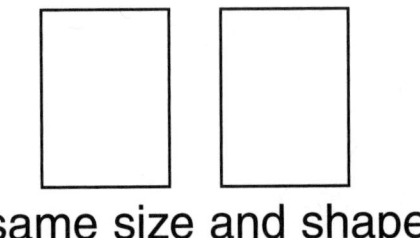

same size and shape

Ring the one that is the same size and shape.

1.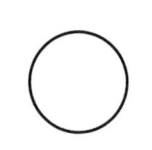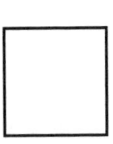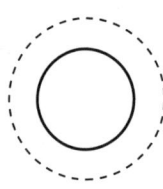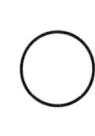

2.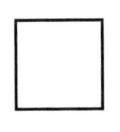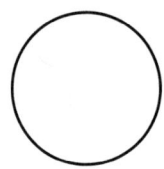

3.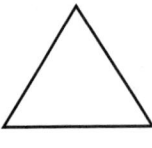

4.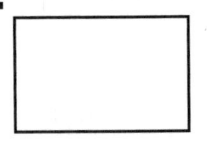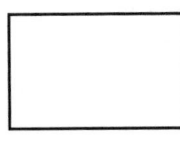

5.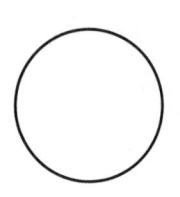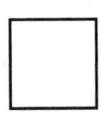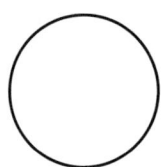

Sort Attribute Blocks by shape.

Matching Size and Shape

Ring the ones that are the same shape and size.

1.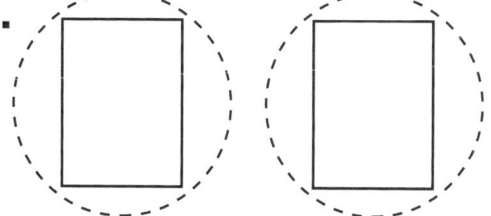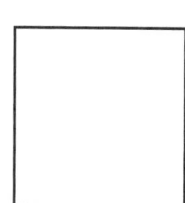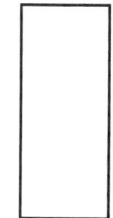

2.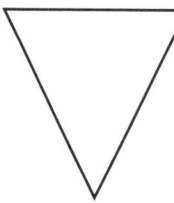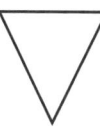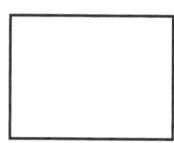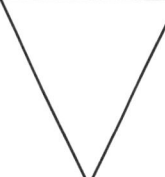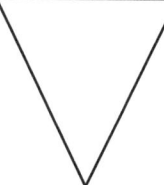

3.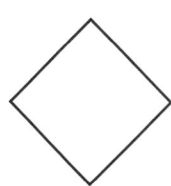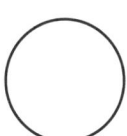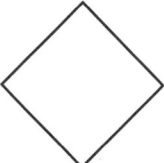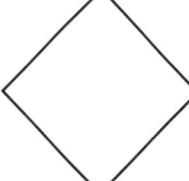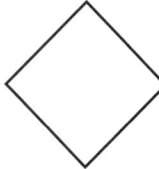

4.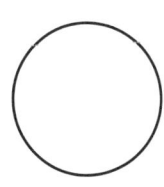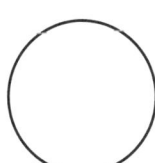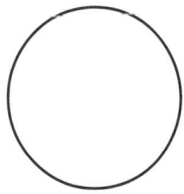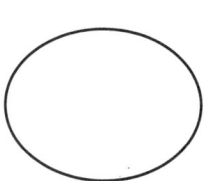

5. Compare the figures to the first one.
Ring the correct word.

 |    

               Same   Same   Same
              Smaller   Smaller   Bigger

Make congruent shapes on Geoboards. ⬭

PLANE FIGURES

Drawing Shapes

Connect the dots.
Draw the shapes.

1.

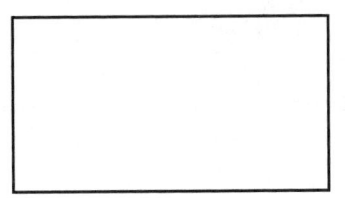

2.

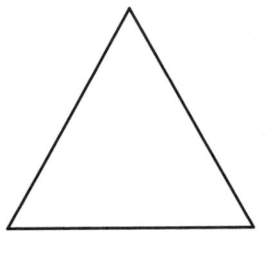 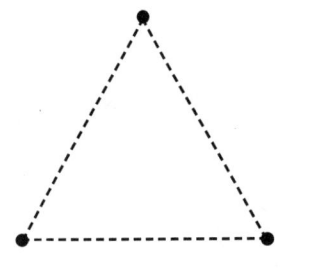 

3.

 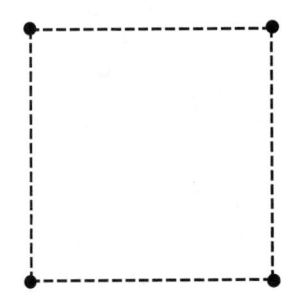

4.

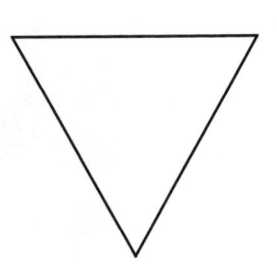 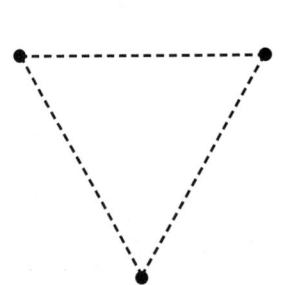

Make shapes on Geoboards.

PLANE FIGURES

Drawing Shapes

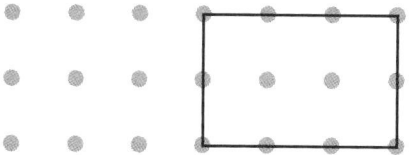 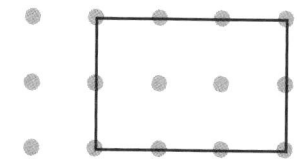

Both figures are the same shape and size.

Copy each figure.
Make yours the same shape and size.

1.

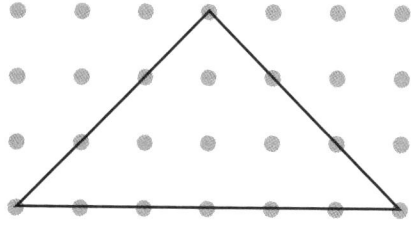

2.

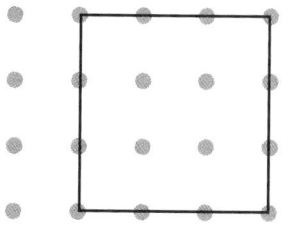

3.

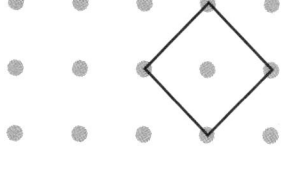

Make shapes on Geoboards.

PLANE FIGURES
Drawing Shapes

1. Connect the dots to make the following shapes.

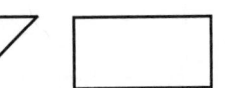

Make 2 of each shape.

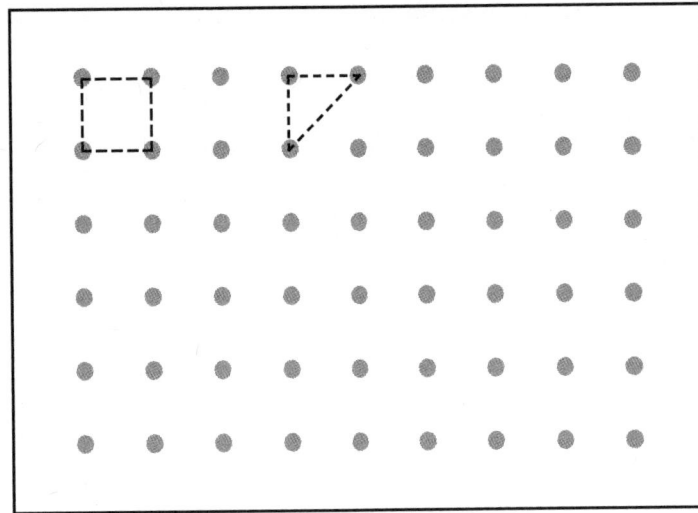

2. Make 7 △ by connecting the dots.
Now make 4 ☐ by connecting the dots.

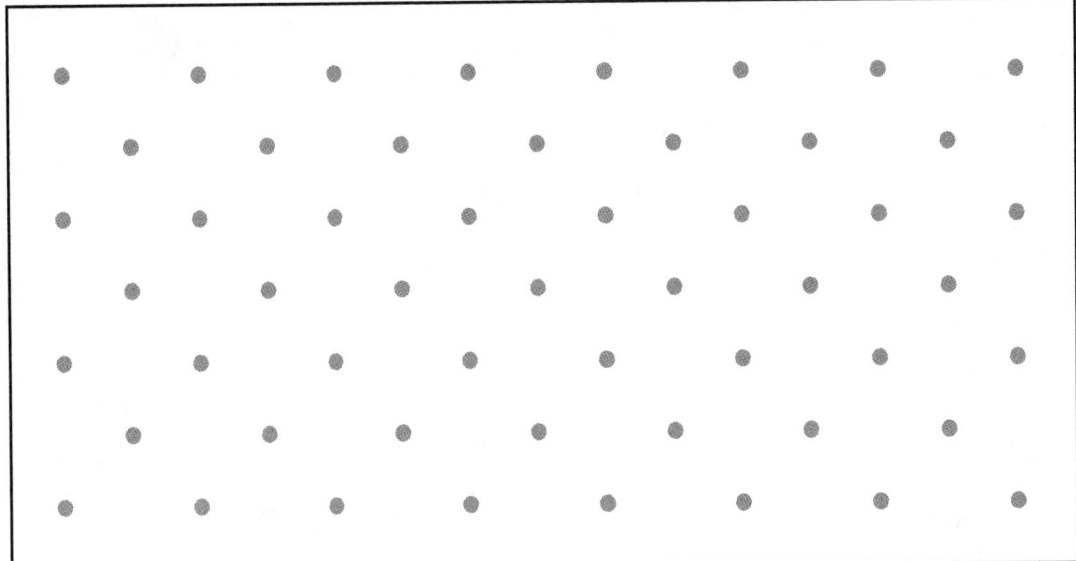

Make shapes on Graph Paper. ◻

PLANE FIGURES
Problem Solving

Count the shapes.
Complete the graph.

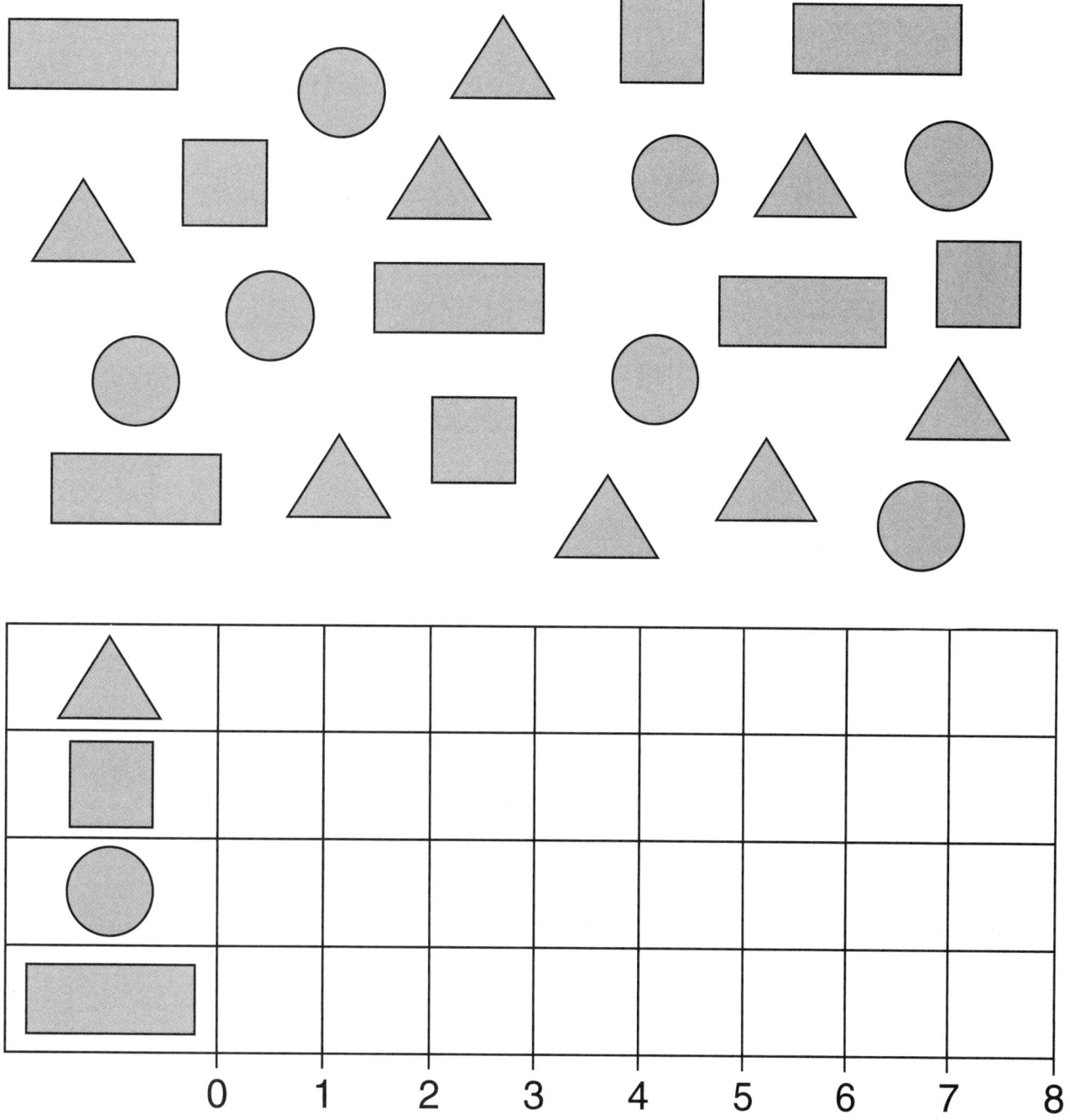

Sort and count Attribute Blocks by shape. ⬭

Problem Solving

Name _____ Date _____

Read each riddle.
Draw the shape.

1. I am a closed figure.
I have 4 sides.
I have 2 long sides.
I have 2 short sides.

2. I am a closed figure.
I have 3 sides.

3. I am a closed figure.
I have 0 sides.
I have 0 corners.

Check answers with Overhead Attribute Blocks. ❓▢

PLANE FIGURES

Problem Solving

Ring the correct picture.

1. Rico drew a triangle.
Then he drew a square.

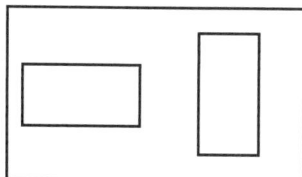 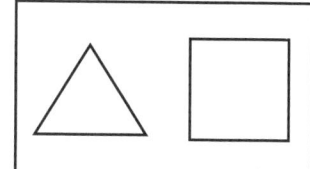

2. Kate drew a circle.
Then she drew a rectangle.

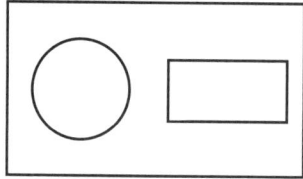 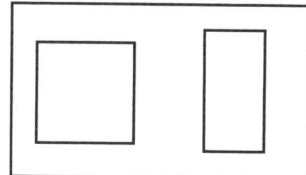

3. Jan drew 2 circles.
They were the same size.

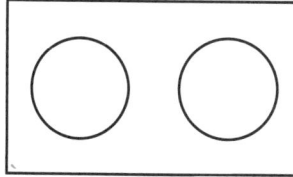 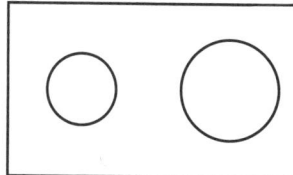

4. Mel drew 2 shapes.
They were the same shape.
They were not the same size.

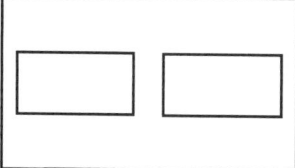 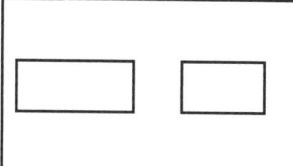

Check answers with Overhead Attribute Blocks. ❓⬜

PLANE FIGURES

Problem Solving

1. Look at the picture.

Find each shape that has six sides.

Color each shape with six sides.

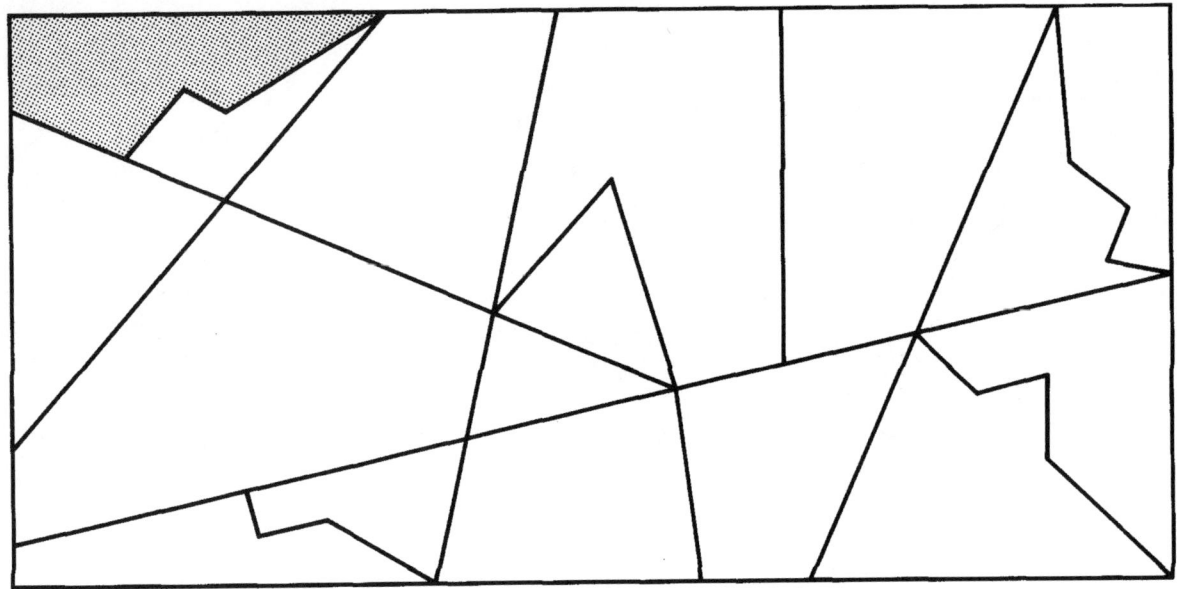

2. Look at the picture.

Find each shape that has eight sides.

Color each shape that has eight sides.

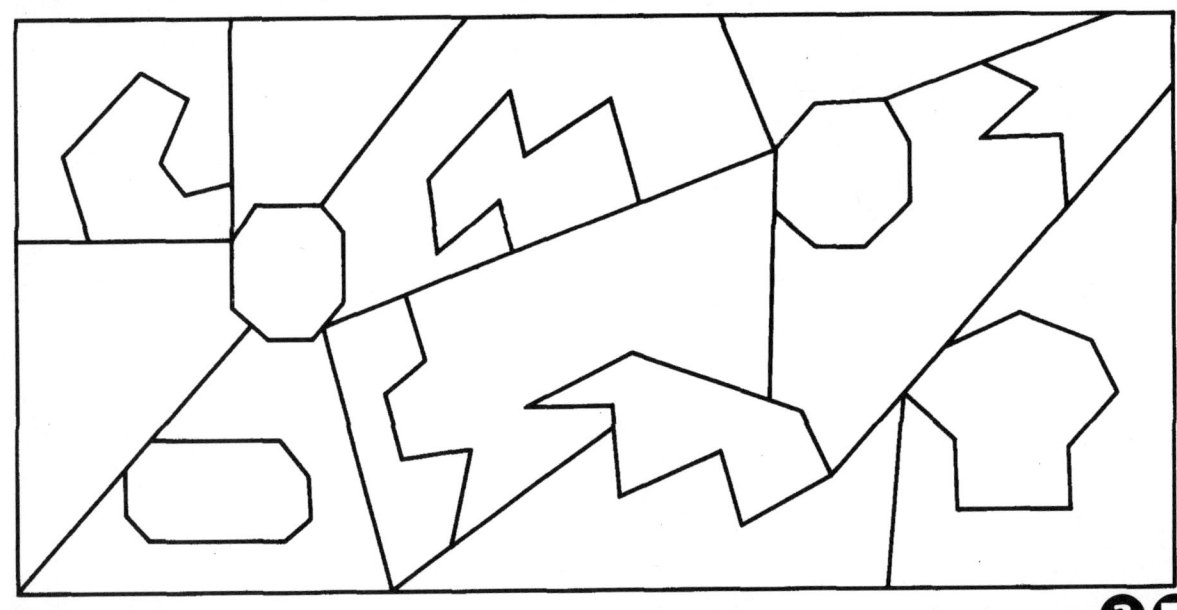

Make six- and eight-sided shapes on Geoboards. ❓⬜

PLANE FIGURES

Problem Solving

1. Add the shape you need to make a .

2. Add the shape you need to make a 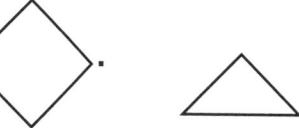.

3. Add the shape you need to make a .

4. Add the shape you need to make a .

5. Now take away the shape you need to make a 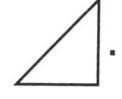.

Color it blue.

6. Take away the shapes you need to make a .

Color them red.

Check answers with Overhead Attribute Blocks. ❓ ⬜

Name _____ Date _____

Assessment: Unit 3

Ring the solid that matches the plane shape.

1.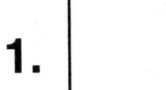

 **a.**   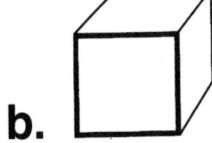 **b.**    **c.**

....

Match the shape.

2.  cylinder

   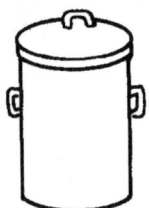

3. Ring the sphere.

      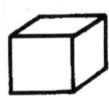

4. Ring the cube.

      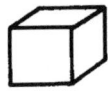

Name _____ Date _____

Matching Plane Shapes to Solids

Match the plane shape to the solid.

1.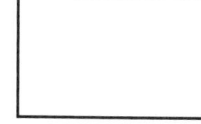

a.

2.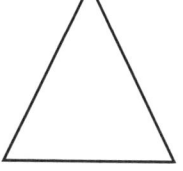

b.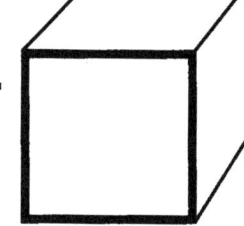

3.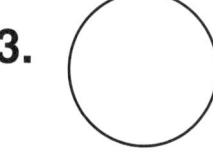

c.

4.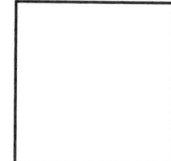

d.

5. Count the △. Write how many.

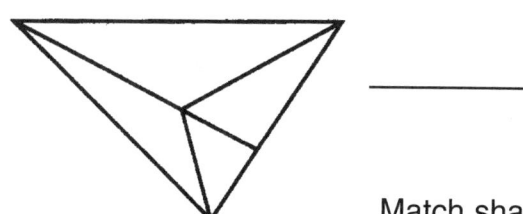 _____

Match shapes to Wooden or Plastic Geometric Solids.

SOLID FIGURES

Matching Plane Shapes to Solids

Match the plane shape to the solid.

1.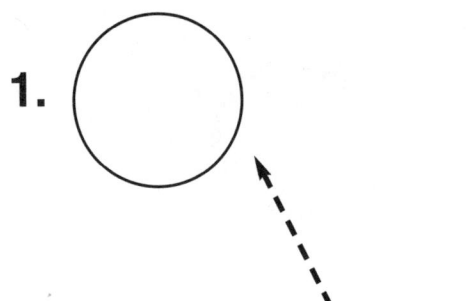

a.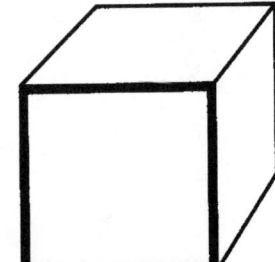

2.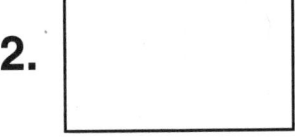

b.

3.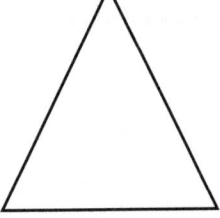

c.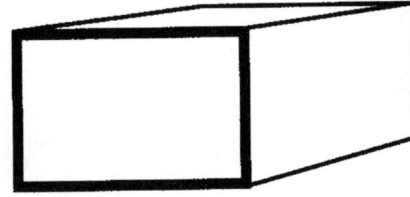

4.

d.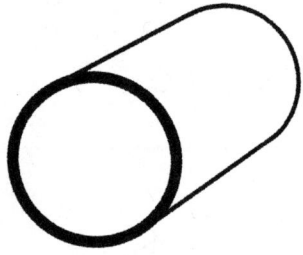

Match shapes to Wooden or Plastic Geometric Solids. ▢

Name _____   Date _____

Identifying Solids

cone

sphere

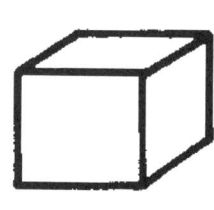
cube

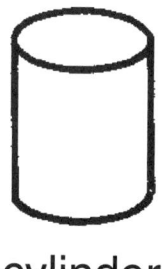
cylinder

Ring the object that has the same shape.

1.
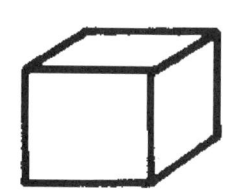         

2.

3.

4.
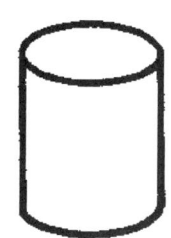         

Match shapes to objects in classroom.

Name _____ Date _____

Identifying Solids

Look at the picture.

1. Color each cone blue.

2. Color each sphere red.

3. Color each cylinder yellow.

4. Color each cube green.

5. Write how many shapes you see.

△		🛢	
○		⬜	

Match shapes to objects in classroom. ⬛

Name _____ Date _____

SOLID FIGURES

Identifying Solids

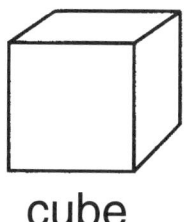 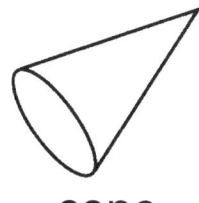 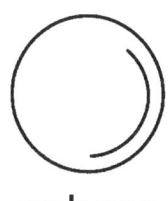 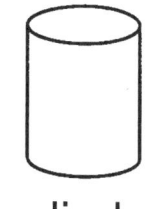 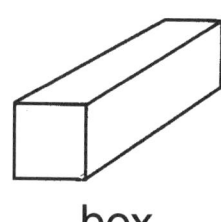

cube   cone   sphere   cylinder   box

1. Color each cube shape red.
2. Color each sphere shape blue.
3. Ring each cylinder shape.
4. Draw an X on each cone shape.
5. Draw a line under each box shape.

Match pictures to Wooden or Plastic Geometric Solids. ⬤

www.svschoolsupply.com
© Steck-Vaughn Company

Unit 3: Solids
Geometry 1, SV 5805-1

SOLID FIGURES

Matching Solids

Draw lines to match.

1.

cube

2.

sphere

3.

cone

4.

cylinder

Match shapes to objects in classroom. ▢

SOLID FIGURES

Identifying Solids

Color each shape the color indicated.

  cone	 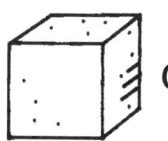 cube	 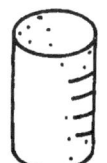 cylinder	  sphere

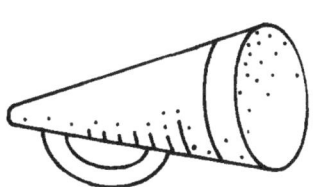

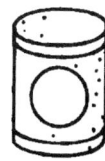

Check answers with Wooden or Plastic Geometric Solids. ▮

Name _____ Date _____

Identifying Solids

Color the  red. Color the  green.

Color the 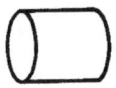 blue. Color the yellow.

1.

Match the shape.

2.    sphere

3.    cylinder

  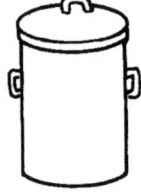

---

4.    cone

5. 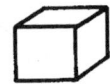   cube

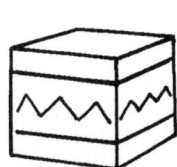  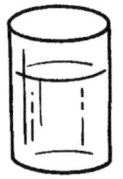

Match shapes to objects in classroom. ▮

SOLID FIGURES

Problem Solving

Count the shapes.
Complete the graph.

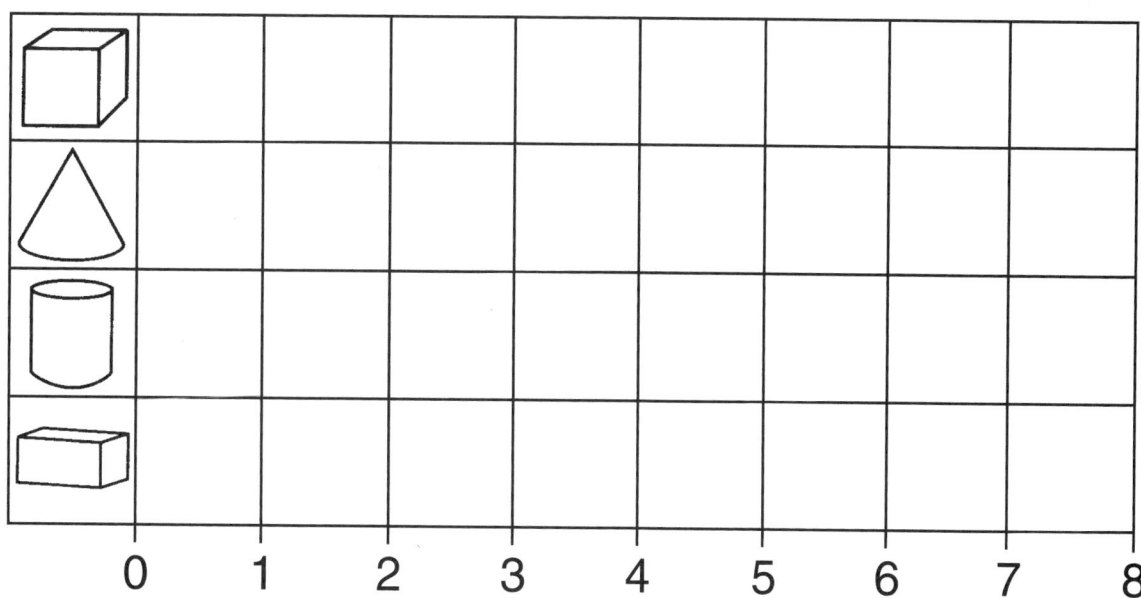

Match pictures to Wooden or Plastic Geometric Solids.

SOLID FIGURES
Problem Solving

Use solid shapes.

Ring each shape that will stack.

1.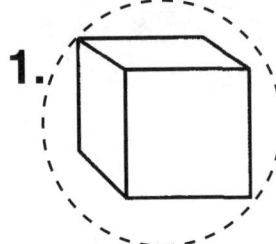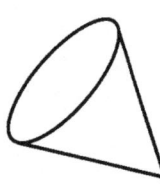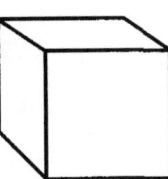

Draw an X on each shape that will roll.

2.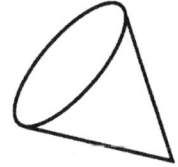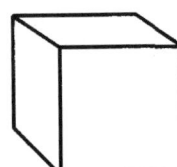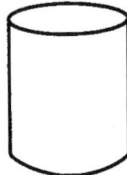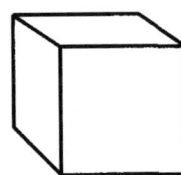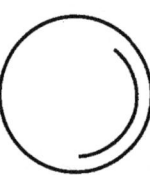

Color each shape that will slide.

3.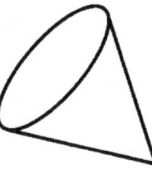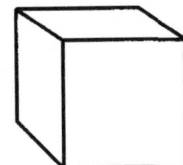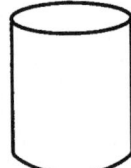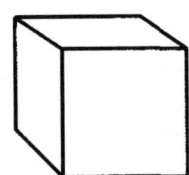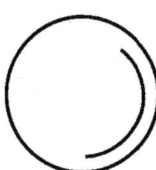

Use solid shapes to build.
Ring the shape that must be on top.

4.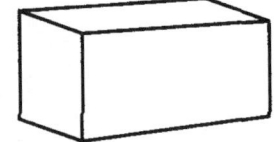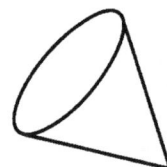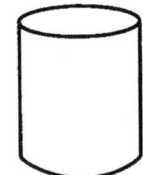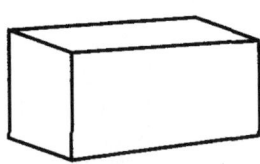

Check answers with Wooden or Plastic Geometric Solids. ❓⬜

Name _____ Date _____

Problem Solving

Use solid shapes.

Color each shape that will stack.

1.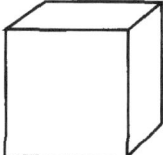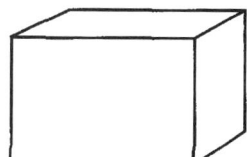

Color each shape that will roll.

2.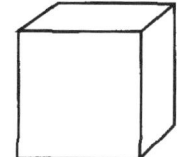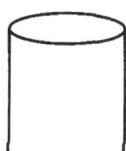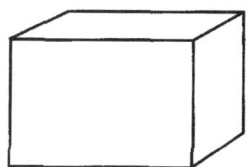

Color each shape that will slide.

3.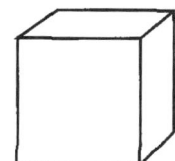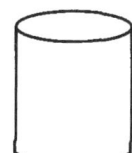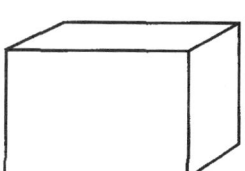

Use solid shapes to build.
Ring each shape that will stack.

4.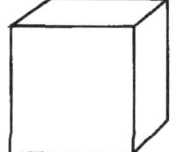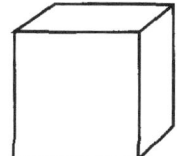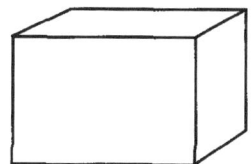

Check answers with Wooden or Plastic Geometric Solids. ❓⬜

Name _____ Date _____

SYMMETRY
Assessment: Unit 4

Do the two parts match?
Ring <u>yes</u> or <u>no</u>.

1.

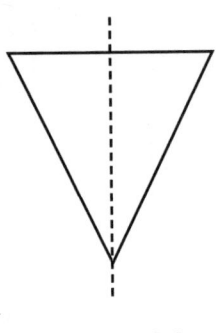

yes   no

2.

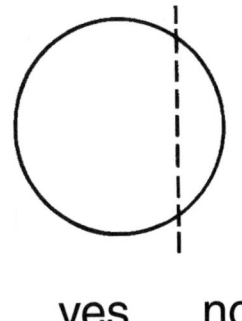

yes   no

3.

yes   no

4.

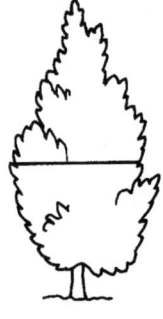

yes   no

5.

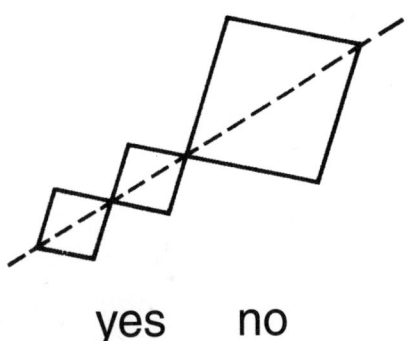

yes   no

6.

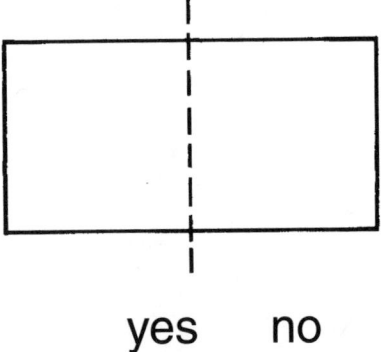

yes   no

SYMMETRY

Identifying Symmetry

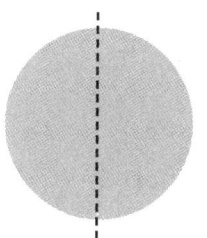

These two parts are the same size and shape.
They match.

....

Do the two parts match?
Ring <u>yes</u> or <u>no</u>.

1.

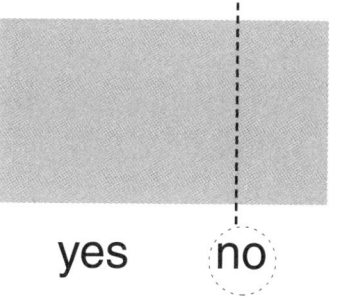

yes   no

2.

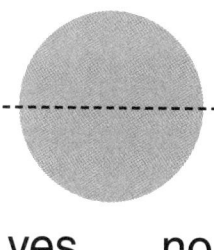

yes   no

3.

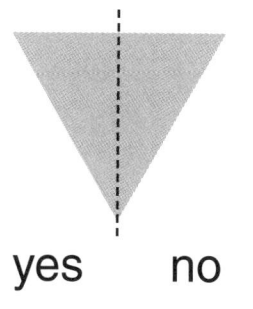

yes   no

4.

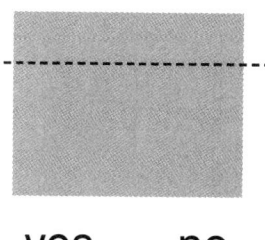

yes   no

5.

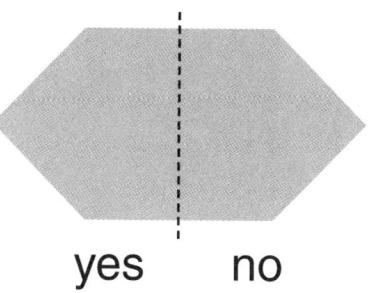

yes   no

6.

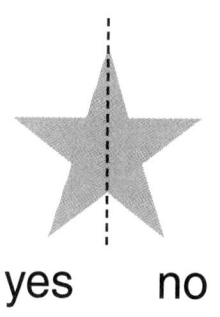

yes   no

Trace shape and fold paper to find line of symmetry. ⬭

SYMMETRY

Drawing Lines of Symmetry

Draw a line to make two parts that match.

1.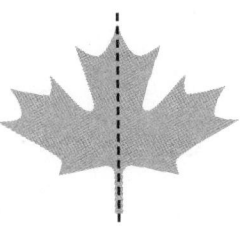

2.

3.

4.

5.

6.

7.

8.

Trace shape and fold paper to find line of symmetry. ▢

Name _____ Date _____

Drawing Lines of Symmetry

Draw a line to make two parts that match.

1.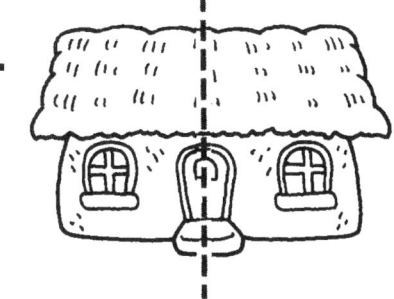

2.

3.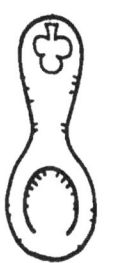

4.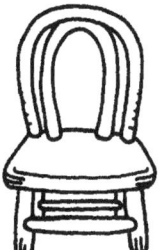

5.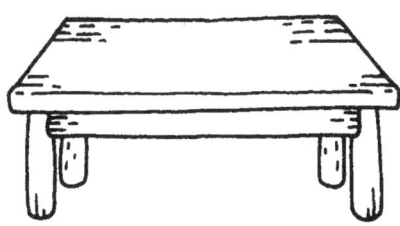

6.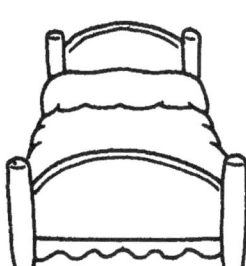

7. Ring the two parts that do **not** match.

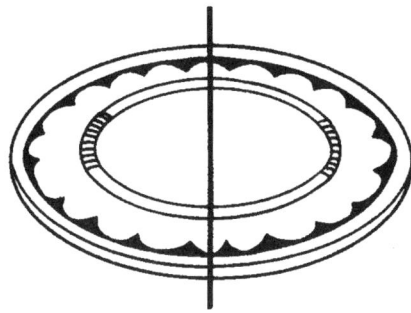  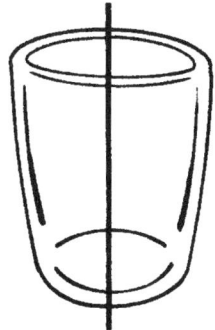

Cut pictures from magazines and fold to find line of symmetry.

Name _____   Date _____

SYMMETRY
Practicing Symmetry

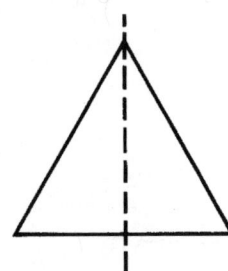 These two parts are the same size and shape.
They match.

....

Do the two parts match? Ring <u>yes</u> or <u>no</u>.

1.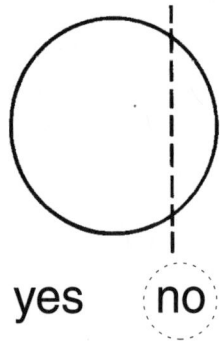

yes   (no)

2.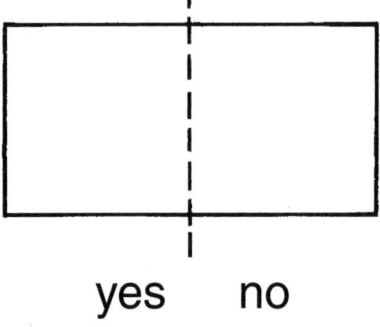

yes   no

3.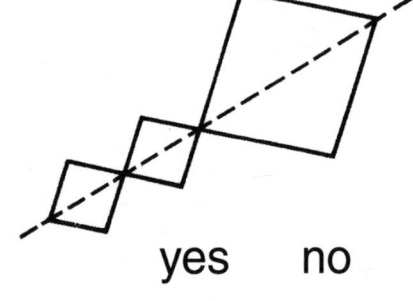

yes   no

4.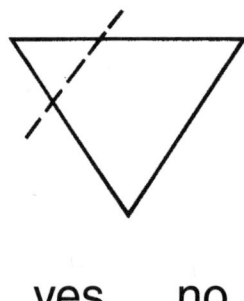

yes   no

....

Draw a line to make two parts that match.

5.

6.

7.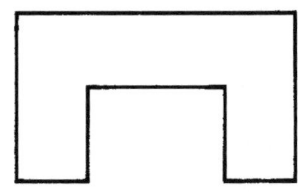

Trace shape and fold paper to find line of symmetry. ⬭

SYMMETRY

Practicing Symmetry

Do the parts match? Ring <u>yes</u> or <u>no</u>.

1.

yes   (no)

2.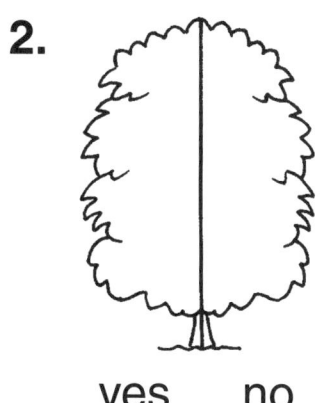

yes   no

3.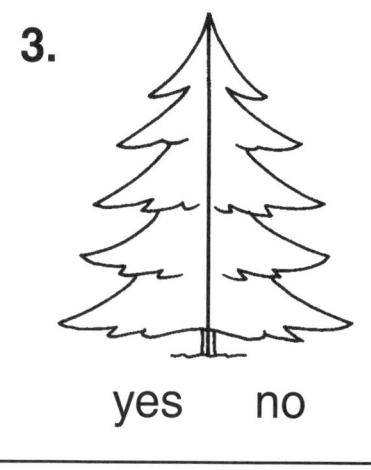

yes   no

4.

yes   no

5.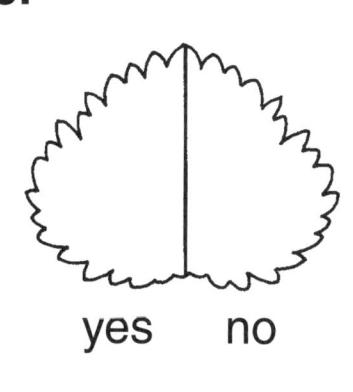

yes   no

6.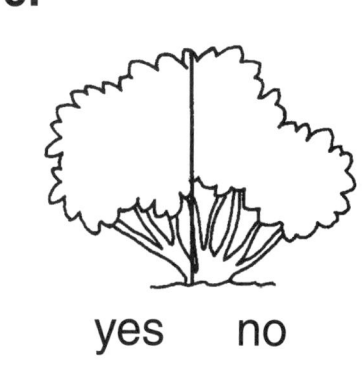

yes   no

Draw a line to make two parts that match.

7.

8.

9.

Trace pictures and fold paper to find line of symmetry. ⬛

MEASUREMENT

Assessment: Unit 5

About how long is each object?

1.
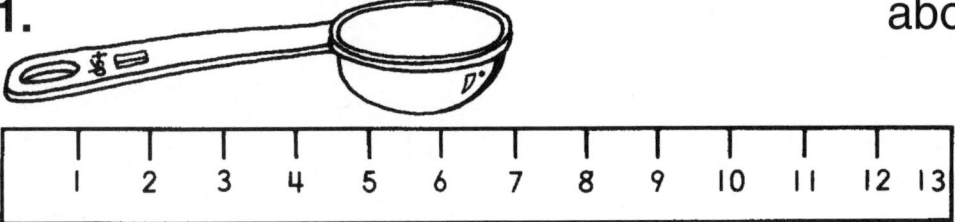
about _____ cm

_____

2.

about _____ inches

....

3. Find how far around.
Measure each side.
Add the lengths.

_____ centimeters

_____ centimeters   _____ centimeters

_____ centimeters in all

....

4. Count the square centimeters.
Write how many in all.

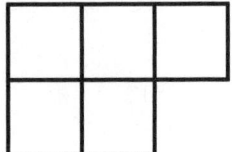   _____ square centimeters

Name _____ Date _____

Measuring Centimeters

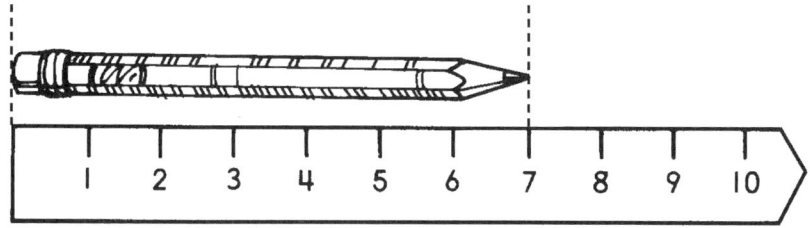

7 centimeters (cm)

How long is each object?

1.

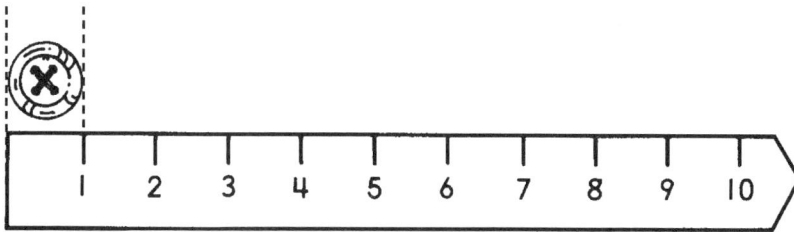

_____ cm

2.

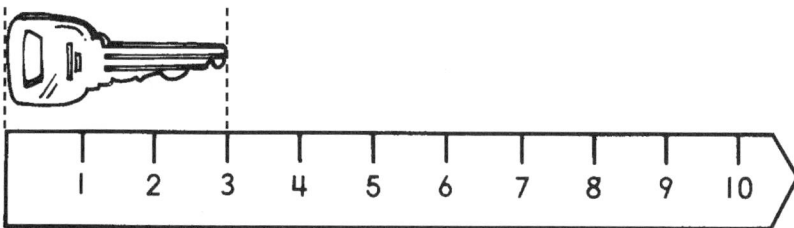

_____ cm

3.

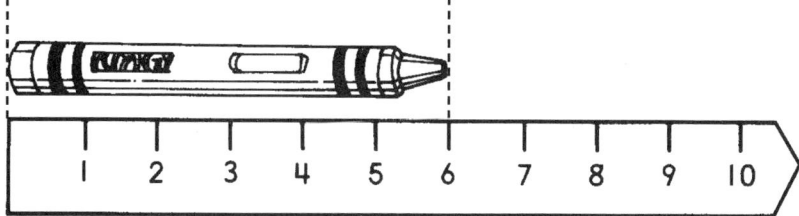

_____ cm

Measure classroom objects with Centimeter Rulers.

Name _____ Date _____

Estimating Centimeters

About how long is each object?

1.

about ___12___ cm

2.

about _____ cm

3. Estimate. Ring the longest.
   Put an X on the shortest.

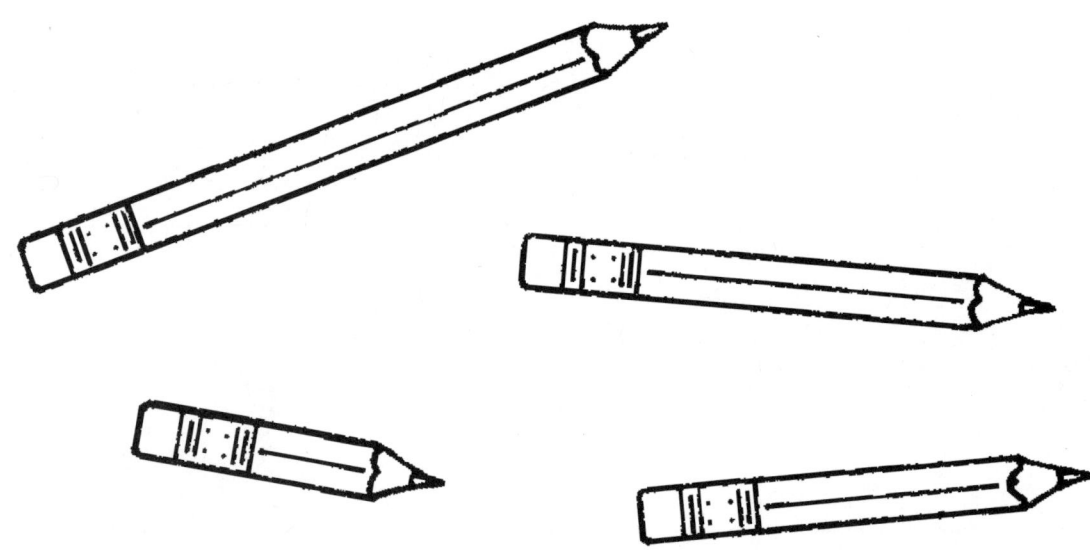

Measure pictures with Centimeter Rulers. ⬤

MEASUREMENT

Measuring Centimeters

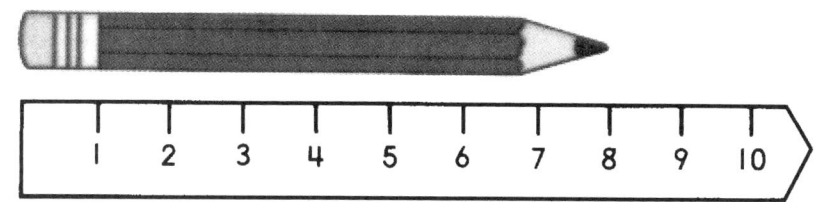

This pencil is 8 cm long.

How long is each pencil?
Use your centimeter ruler.

1.

_____ cm

2.

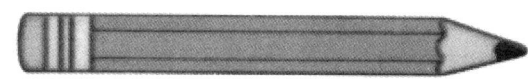

_____ cm

3.

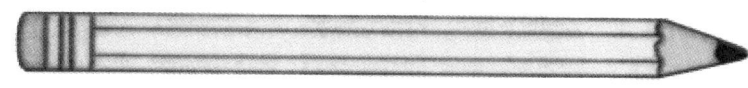

_____ cm

4.

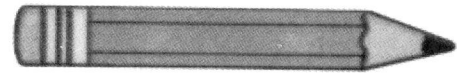

_____ cm

5.

_____ cm

Measure pictures with Centimeter Rulers.

MEASUREMENT

Measuring Centimeters

About how long is each object?

1.   about ___13___ cm

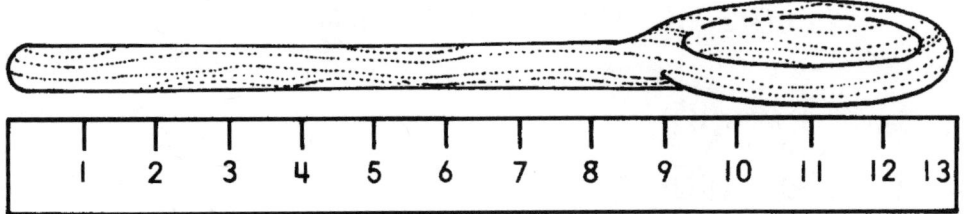

2.   about _____ cm

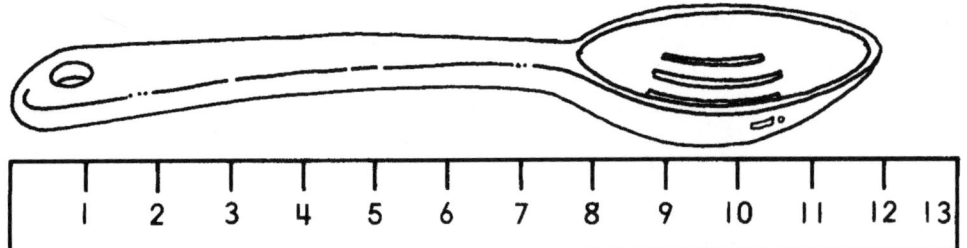

3.   about _____ cm

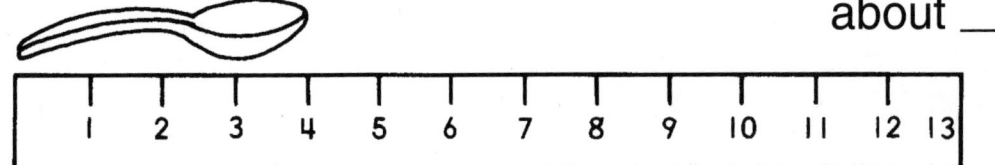

4.   about _____ cm

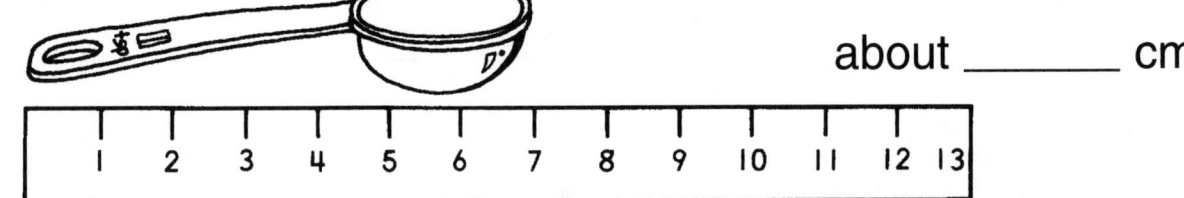

Measure classroom objects with Centimeter Rulers. ⬤

Name _____ Date _____

Measuring Centimeters

Write the number of centimeters.

1.

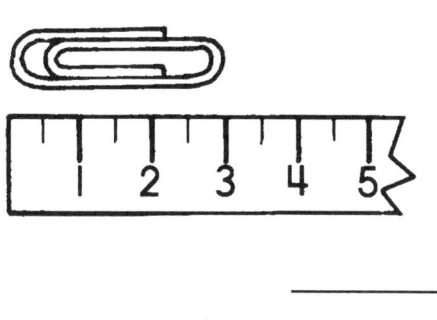

_____

2.

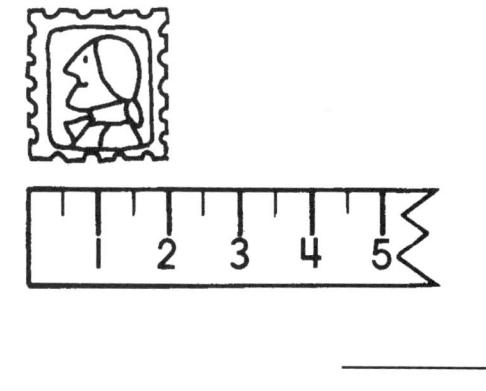

_____

About how long is each object?

3.

about _____ cm

4.

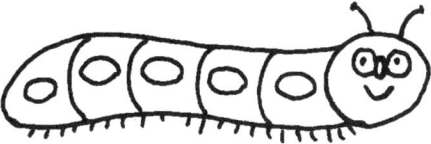

about _____ cm

5.

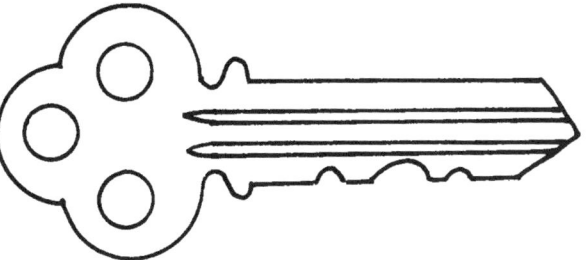

about _____ cm

Measure classroom objects with Centimeter Rulers.

Name _____ Date _____

Measuring Inches

How long is the feather?

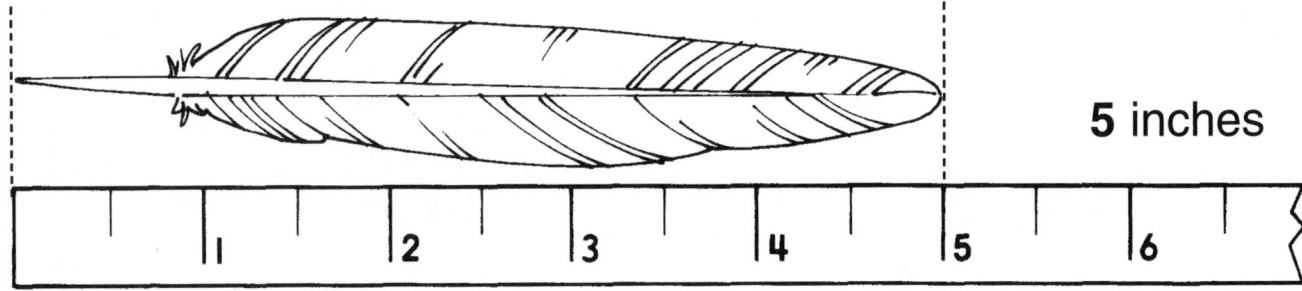

5 inches

....

How long is each feather?

1.

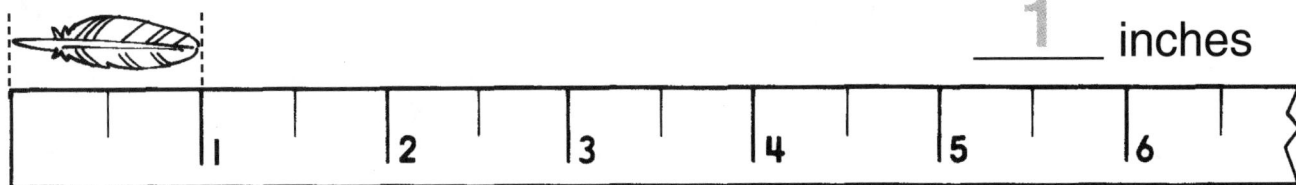

_1_ inches

2.

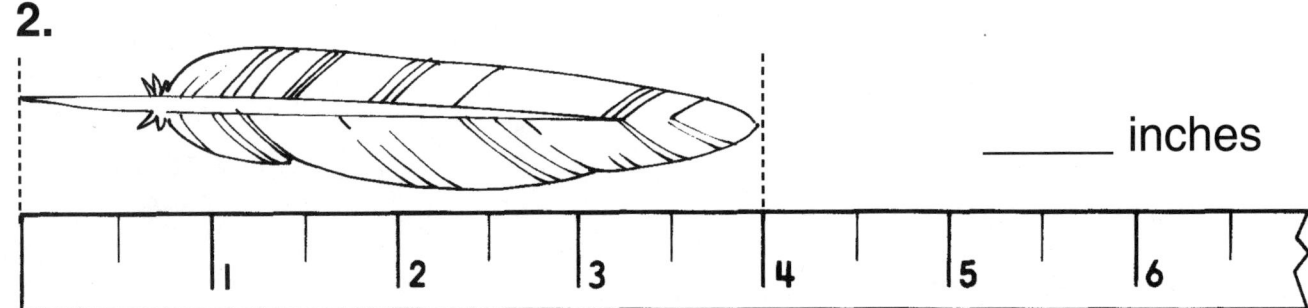

_____ inches

3.

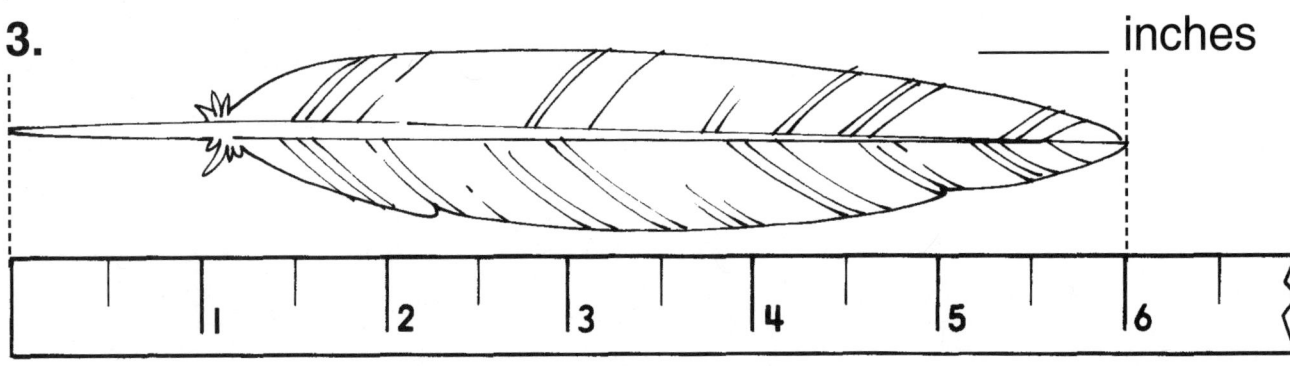

_____ inches

Measure outdoor objects with Inch Rulers.

MEASUREMENT

Measuring Inches

How long is each leaf?

1.

about __3__ inches

2.

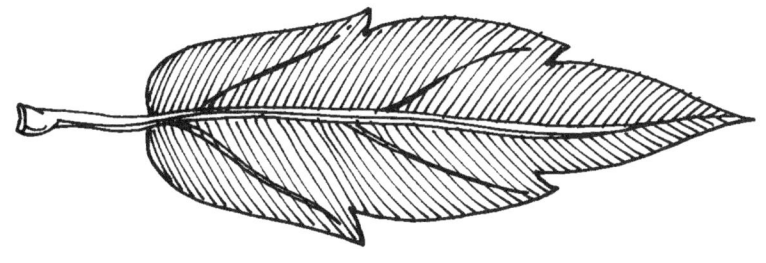

about _____ inches

3.

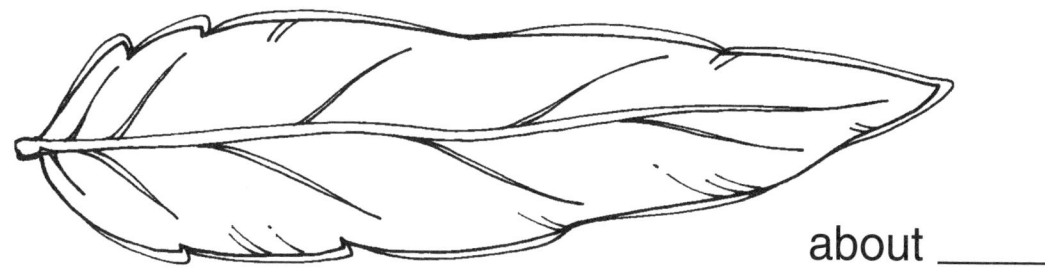

about _____ inches

4.

about _____ inches

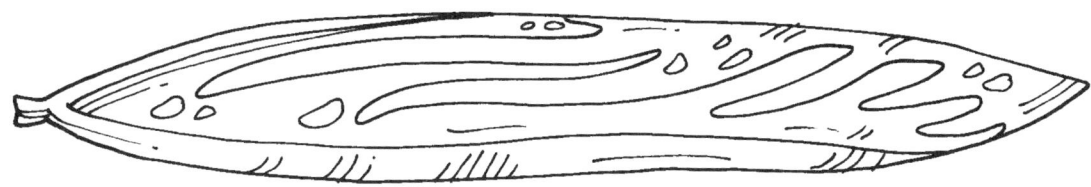

Measure outdoor objects with Inch Rulers.

Name _____ Date _____

Measuring Inches

How long is each carrot?

1.

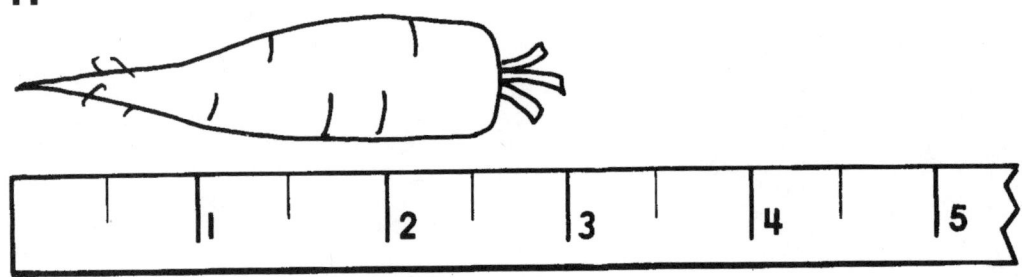

3
_____
inches
long

2.

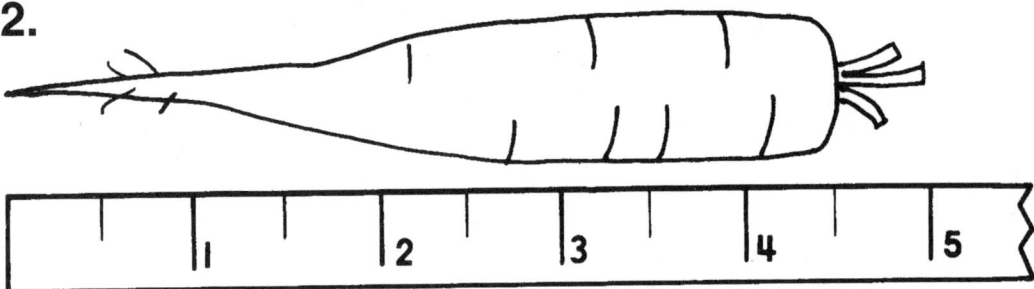

_____
inches
long

....

How long is each ear of corn?

3.

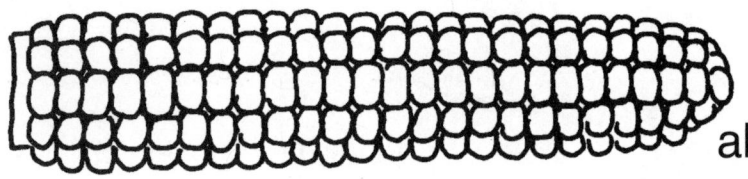

about _____ inches long

4.

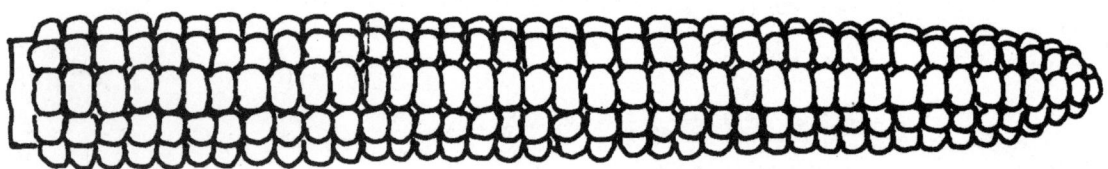

about _____ inches long

Measure classroom objects with Inch Rulers.

Name _____ Date _____

Measuring Inches

How long is each object?

1.

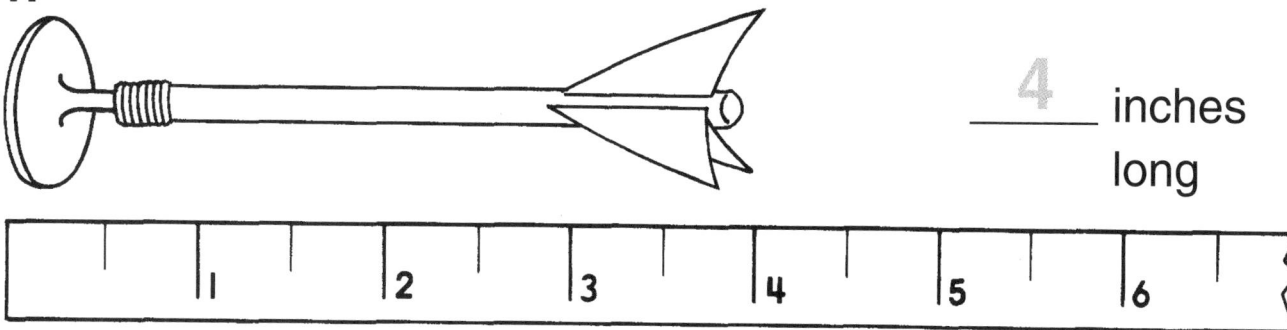

4 ___ inches long

2.

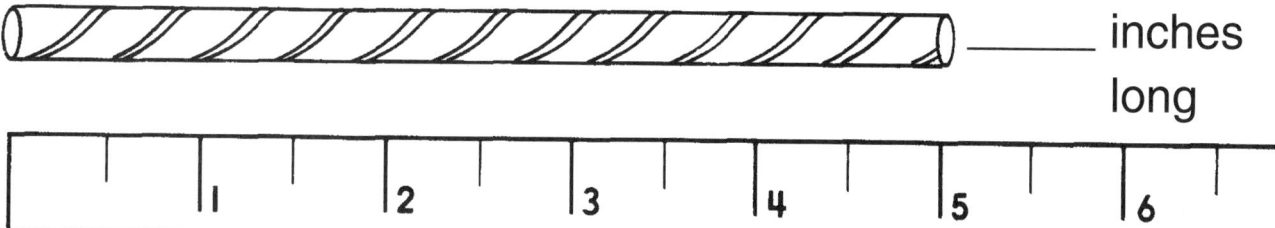

_____ inches long

3.

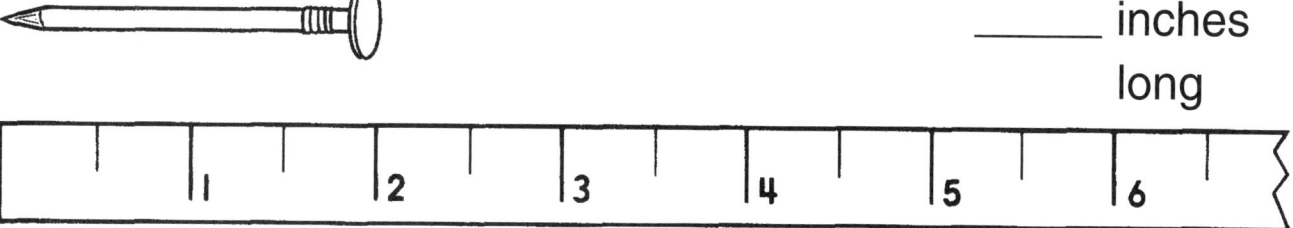

_____ inches long

4.

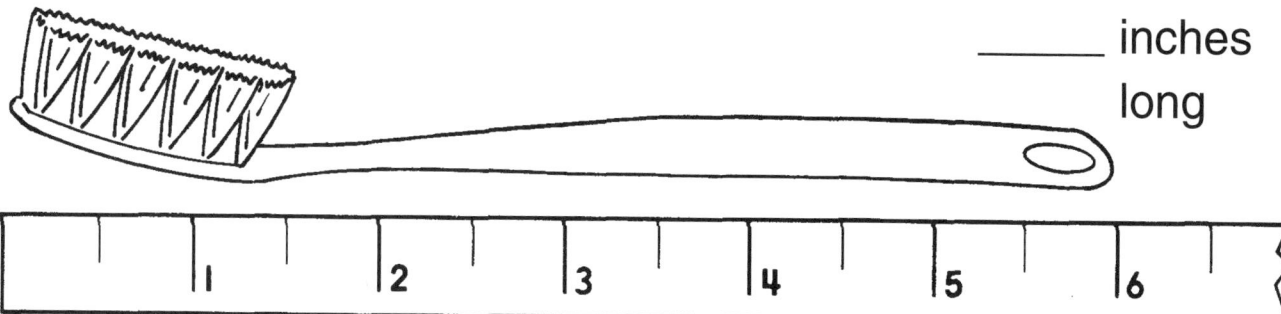

_____ inches long

Measure classroom objects with Inch Rulers.

MEASUREMENT

Drawing Shapes

Use your own ruler to draw each shape.

1. Draw a ☐. Make each side 2 cm long.

2. Draw a ▭. Make two sides 3 cm long.
Make the other two sides 2 cm long.

3. Draw a ▭. Make two sides 4 cm long.
Make the other two sides 2 cm long.

....

4. Now draw a Zany Zoid.
Follow these measurements.
Two sides are 2 cm long.
Three sides are 4 cm long.

Check work with Centimeter Rulers. ◻ €

Name _____ Date _____

MEASUREMENT

Measuring Perimeter

Find how far around.
Measure each side.
Add the lengths.

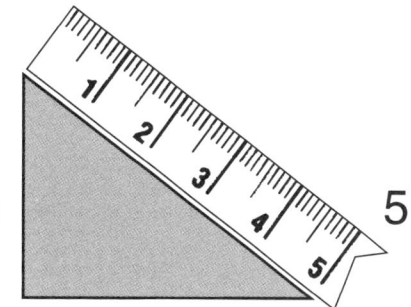

3 centimeters   5 centimeters

4 centimeters

3 + 4 + 5 = 12 centimeters

....

Measure each side.
Then add to find how far around.

1.   ___2___ centimeters 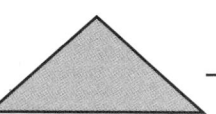 ___2___ centimeters

___3___ centimeters

___7___ centimeters in all

_____

2.   _____ centimeters

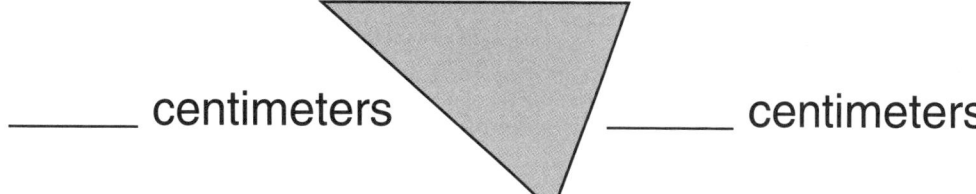

_____ centimeters   _____ centimeters

_____ centimeters in all

Check addition with Calculator. ◻▤

MEASUREMENT
Measuring Perimeter

How many centimeters around does each shape measure?

Add with
a calculator. ___4___ + ___9___ + ___4___ + ___9___ = ___26___ centimeters

PRESS **ON/C** 4 **+** 9 **+** 4 **+** 9 **=** *26.*

Use a centimeter ruler.
Measure each side.
Add with a calculator.

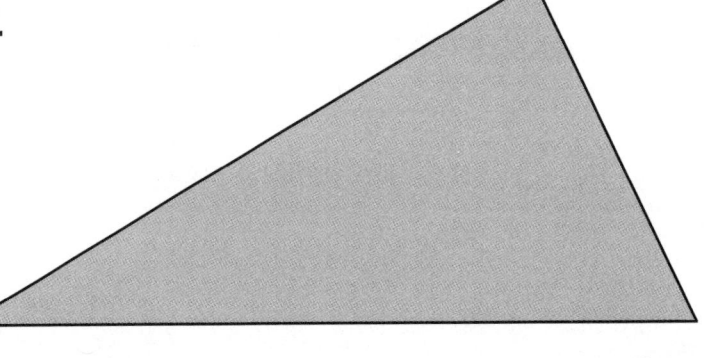

1. ____ + ____ + ____ = ____ centimeters

2. ____ + ____ + ____ + ____ = ____ centimeters

Check addition with Calculator. ⬤ €

Name _____ Date _____

Measuring Area

1 square centimeter

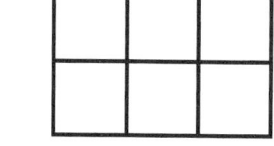

6 square centimeters

Count the square centimeters.
Write how many in all.

1.

___2___ square centimeters

2.

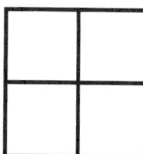

_____ square centimeters

3.

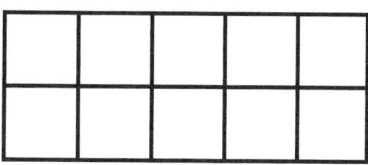

_____ square centimeters

4.

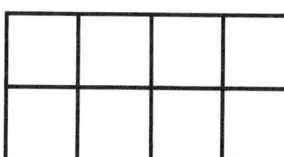

_____ square centimeters

5.

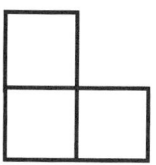

_____ square centimeters

6.

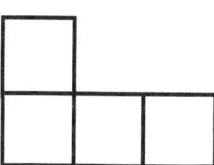

_____ square centimeters

Make other problems on Centimeter Graph Paper.

Measuring Perimeter and Area

Measure each side. Then add to find how far around.

1.

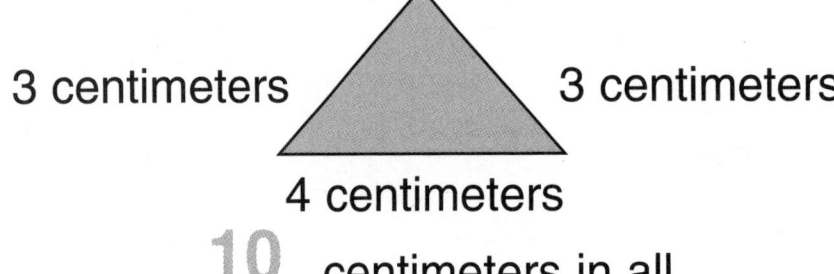

3 centimeters   3 centimeters

4 centimeters

_____10_____ centimeters in all

2.

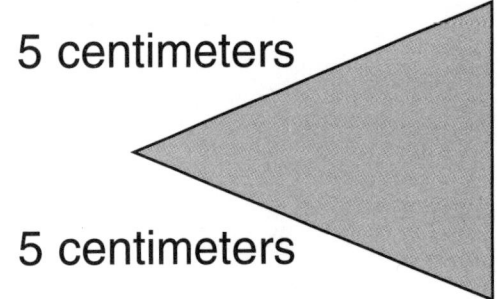

5 centimeters

4 centimeters

5 centimeters   _____ centimeters in all

Count the square centimeters.
Write how many in all.

3.

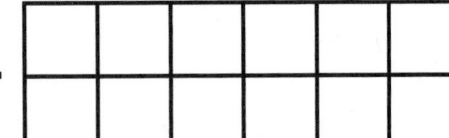

_____ square centimeters

4.

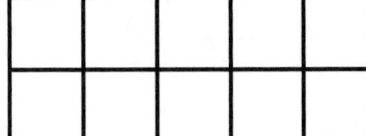

_____ square centimeters

5.

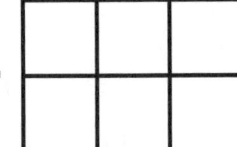

_____ square centimeters

6.

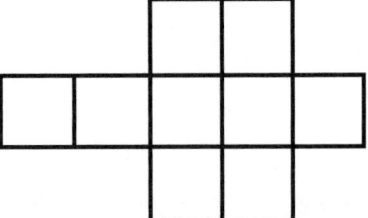

_____ square centimeters

Make other problems on Centimeter Graph Paper.

M E A S U R E M E N T

Problem Solving

Draw a different shape.
Use the same number of square centimeters.

1.

2.

3.

4.

Make other shapes on Centimeter Graph Paper.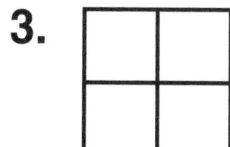

Name _____ Date _____

Assessment: Unit 6

Ring the shape to continue the pattern.

1.

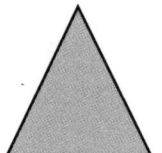     

2.

      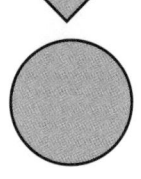

Read each pattern.
Draw the missing shape.

3.

 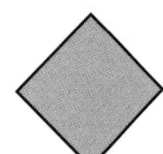 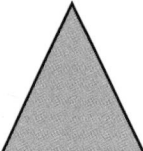 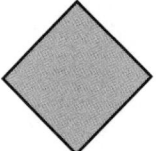  _____

4.

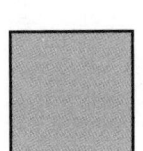   _____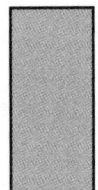

PATTERNS

Completing Patterns

Ring the shape to continue the pattern.

1.

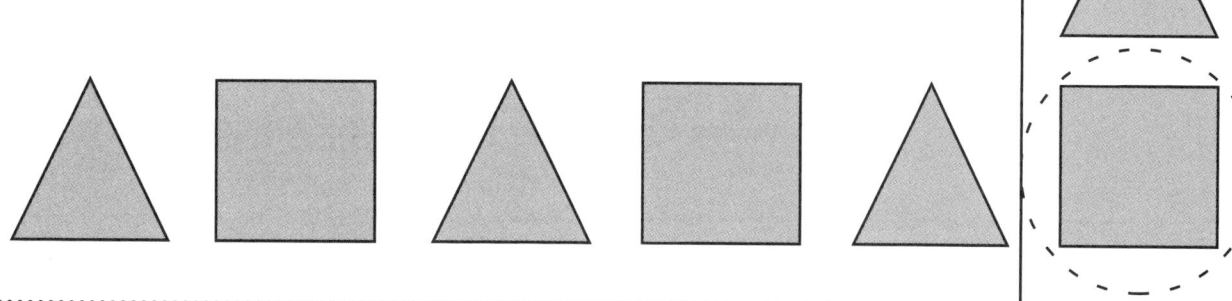

2.

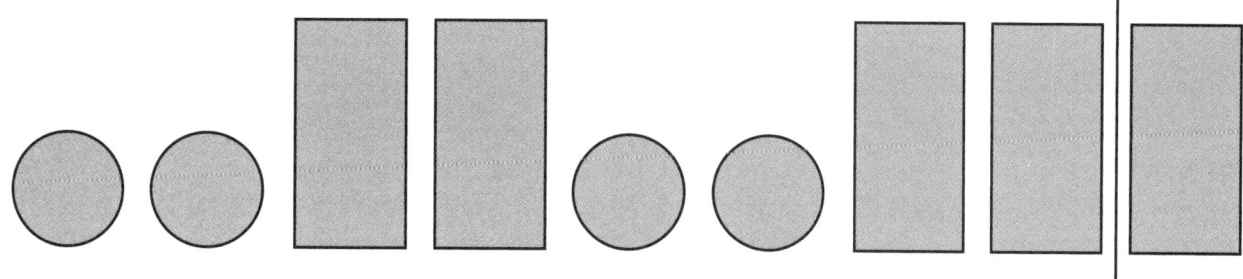

3.

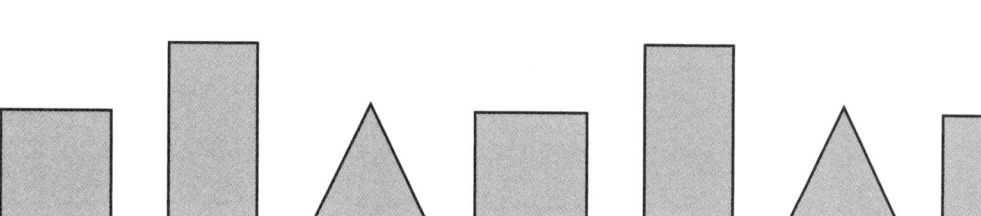

Duplicate patterns using Attribute Beads.

Name _____ Date _____

Completing Patterns

Ring the shape to continue the pattern.

1.

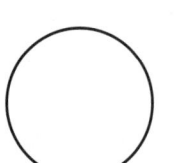 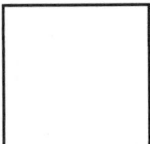 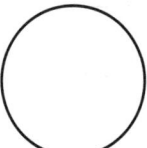 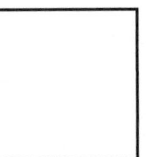 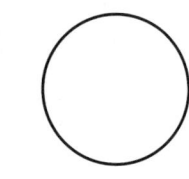 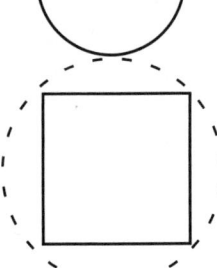

2.

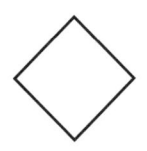 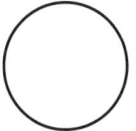 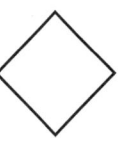 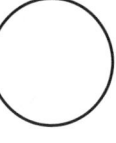 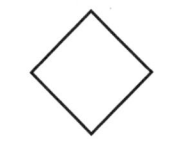 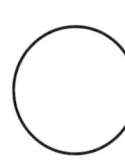 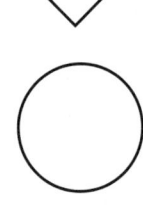

3.

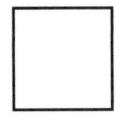    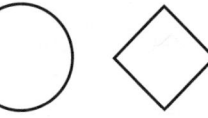  

4.

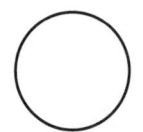   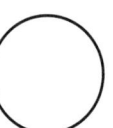 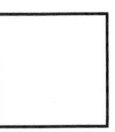 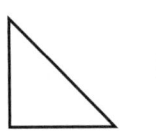 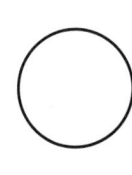 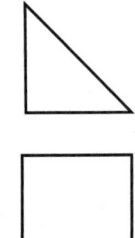

Duplicate patterns using Attribute Beads.

Name _____   Date _____

Completing Patterns

Draw the shape and write the number that comes next.

1.

2.

3.

    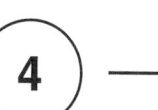

4.

    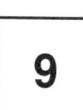 

Duplicate patterns using Attribute Beads.

PATTERNS

Completing Patterns

Match to continue each pattern.

1. ● ● ○ ● ● ○ ● ● ○

a. ○ ○ ● ○ ○ ●

b. ● ● ○ ● ● ○

c. ● ○ ● ○ ● ○

2. ○ ● ● ○ ● ● ○ ● ●

a. ○ ● ● ○ ● ●

b. ● ● ○ ● ● ○

c. ● ● ○ ● ● ○

3. ● ● ● ● ● ● ● ● ●

a. ● ● ● ● ● ●

b. ● ● ● ● ● ●

c. ● ● ● ● ● ●

4. ● ○ ○ ● ● ● ● ○ ○

a. ● ● ● ○ ○ ○

b. ● ● ● ● ○ ○

c. ○ ○ ● ● ● ●

Duplicate patterns using Sorting Beads. ▢ ⊙

PATTERNS

Problem Solving

Read each pattern.
Draw the missing shape.

1.

2.

   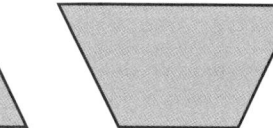  ____

3.

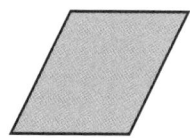  ____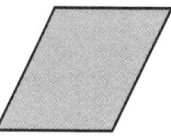

4.

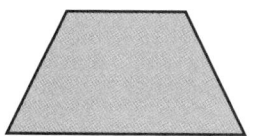   ____

5.

  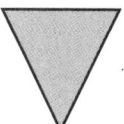  ____

Duplicate patterns using Attribute Lacing Buttons.

Name _____   Date _____

Assessment: Unit 7

1. Ring each shape that shows $\frac{1}{2}$.

 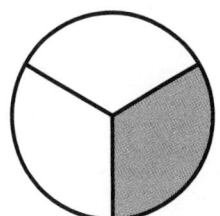 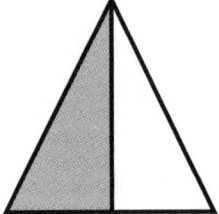

2. Ring each shape that shows $\frac{1}{3}$.

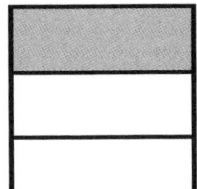 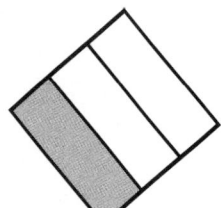 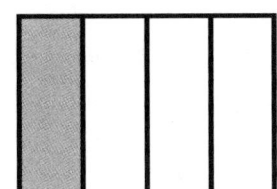

3. Ring each shape that shows $\frac{1}{4}$.

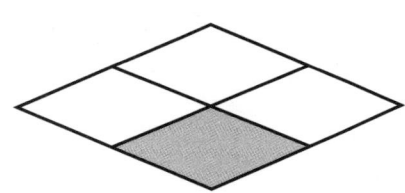 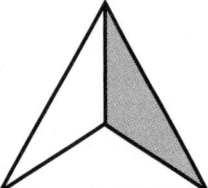 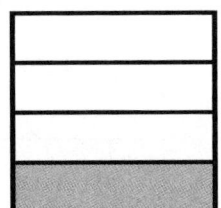

Ring the correct fraction.

4.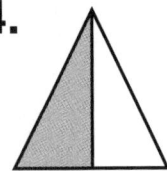

$\frac{1}{2}$ $\frac{1}{3}$ $\frac{1}{4}$

5.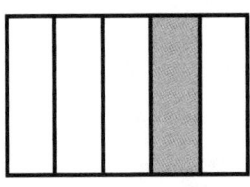

$\frac{1}{3}$ $\frac{1}{4}$ $\frac{1}{5}$

6.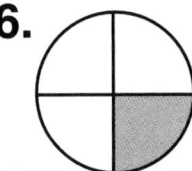

$\frac{1}{3}$ $\frac{1}{4}$ $\frac{1}{5}$

Unit 7: Fractions Using Pictorial Models
Geometry 1, SV 5805-1

Identifying Equal Parts

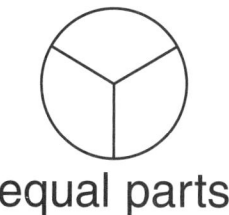
equal parts

not equal parts

Color the shape that shows equal parts.

1.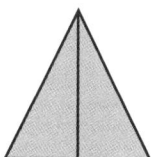

2.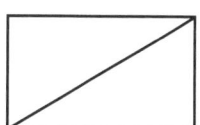

3.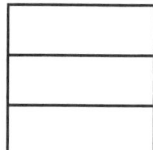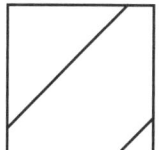

4.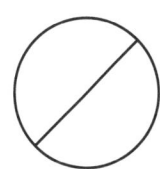

Trace the lines that make equal parts.
Ring each shape that has equal parts.

5.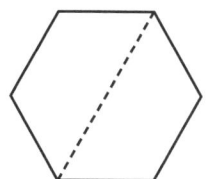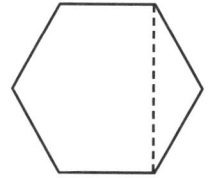

6.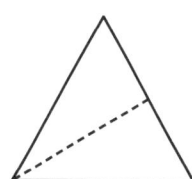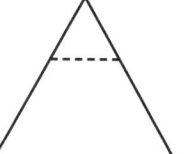

Demonstrate equal parts with Overhead Fraction Circles, Squares, and Rectangles.

Name _____ Date _____

Identifying Equal Parts

equal parts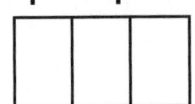

not equal parts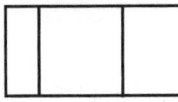

....

Ring the shape that shows equal parts.

1.

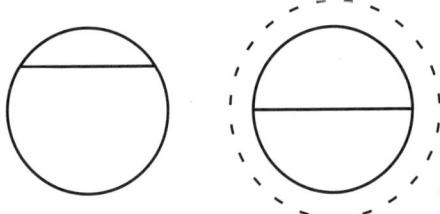

2.

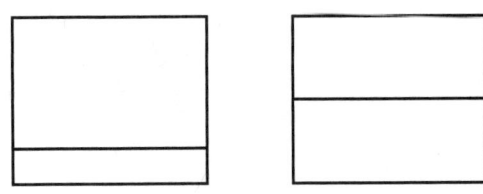

3.

4.

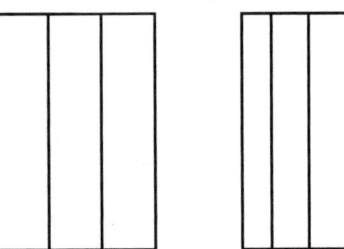

5.

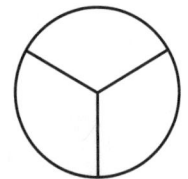 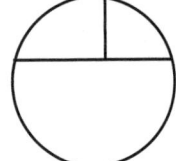

6.

 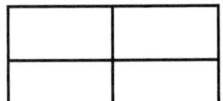

Demonstrate equal parts with Overhead Fraction Circles, Squares, and Rectangles. ◗

Identifying Equal Parts

Does the shape have equal parts?
Ring <u>yes</u> or <u>no</u>.

1.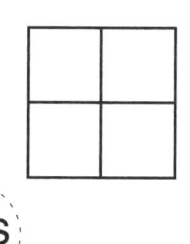

yes   no

2.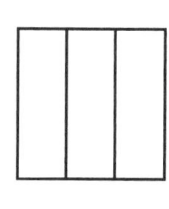

yes   no

3.

yes   no

4.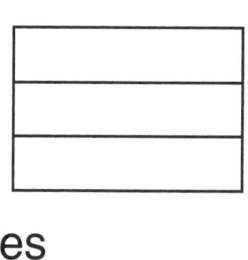

yes   no

5.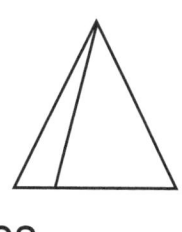

yes   no

6.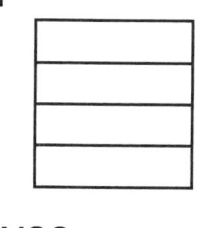

yes   no

Color the shapes with equal parts.

7.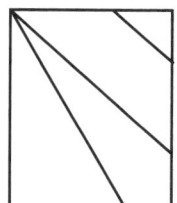

8.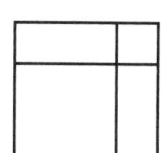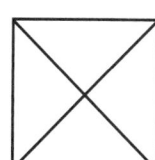

9.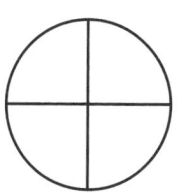

10.

Duplicate equal parts with Overhead Fraction Circles, Squares, and Rectangles.

FRACTIONS USING PICTORIAL MODELS

Identifying Equal Parts

Trace the lines that make equal parts.
Color the shapes that show equal parts.

1.

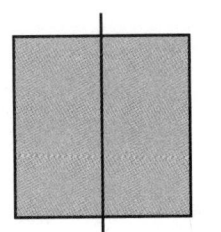 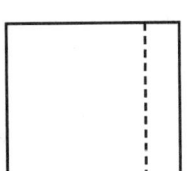

2.

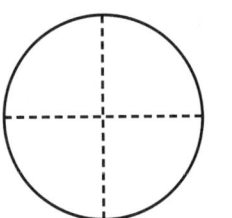 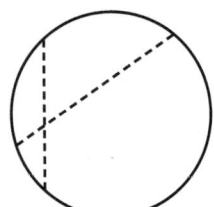

3.

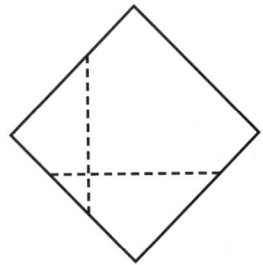 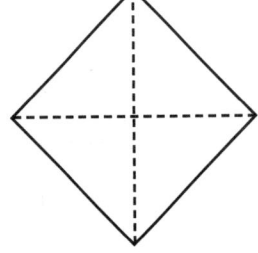

4.

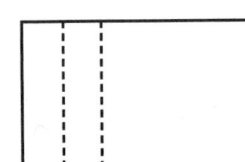 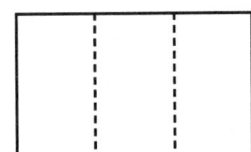

5.

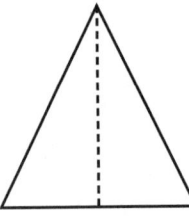 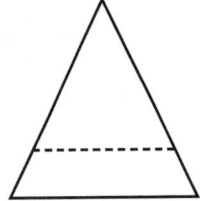

6.

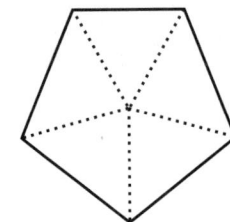 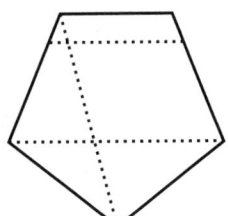

Duplicate equal parts with Overhead Fraction Circles, Squares, and Rectangles.

Name _____ Date _____

Dividing into Equal Parts

Show another way to cut the food
into the same number of equal parts.
Make the pieces the same size.

Trace and cut pictures into equal parts. ❓ ⬭ ▶

Identifying One Half

Draw a line to show two equal parts.
Then color to show $\frac{1}{2}$.

1.

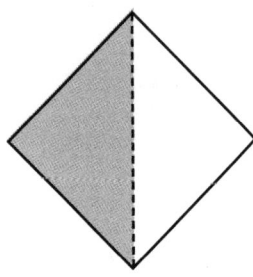

2.

3.

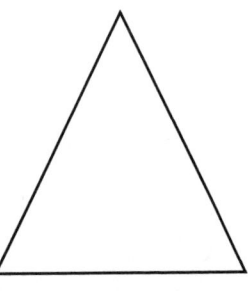

4.

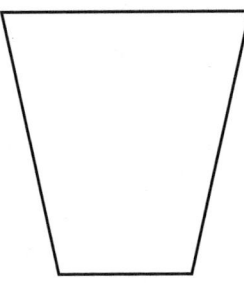

Find the shapes that show two equal parts. Color $\frac{1}{2}$.

5.

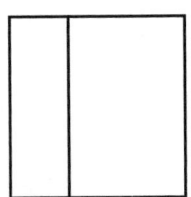 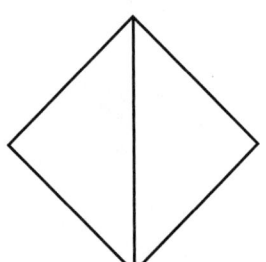

Demonstrate fractions with Overhead Fraction Circles.

Identifying One Third

Ring the fraction that each shape shows.

1.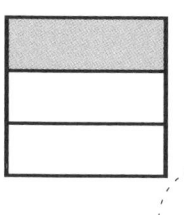

$\frac{1}{2}$ $\frac{1}{3}$

2.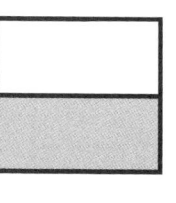

$\frac{1}{2}$ $\frac{1}{3}$

3.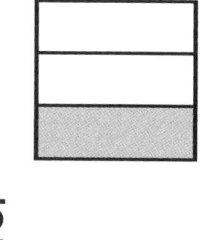

$\frac{1}{2}$ $\frac{1}{3}$

4.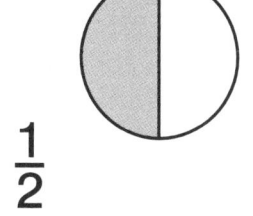

$\frac{1}{2}$ $\frac{1}{3}$

5.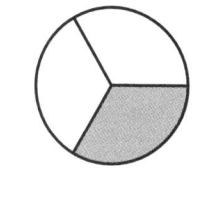

$\frac{1}{2}$ $\frac{1}{3}$

6.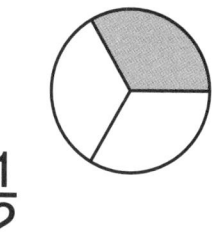

$\frac{1}{2}$ $\frac{1}{3}$

Find the shapes that show three equal parts. Color $\frac{1}{3}$.

7.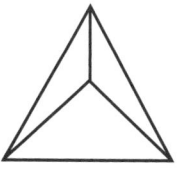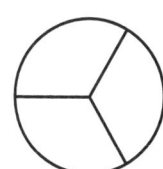

8.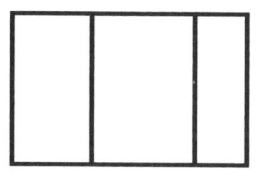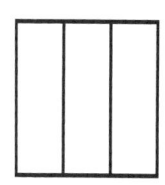

9. Which is greater, $\frac{1}{2}$ or $\frac{1}{3}$?

Ring the greater fraction.

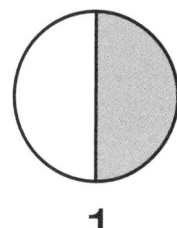

$\frac{1}{2}$ $\frac{1}{3}$

Duplicate fractions with Fraction Builder Strips.

Name _____ Date _____

FRACTIONS USING PICTORIAL MODELS

Identifying One Half and One Third

2 equal parts

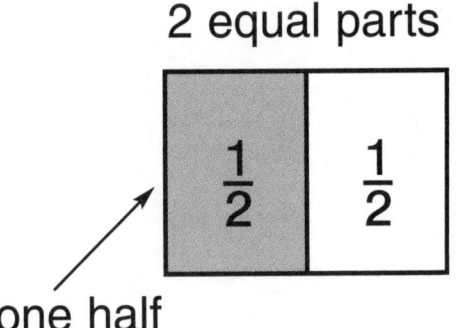

one half

3 equal parts

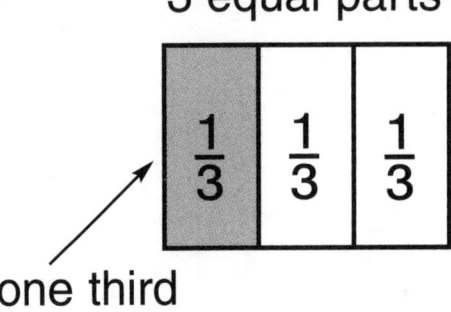

one third

Divide the shapes to show the fraction.

Show $\frac{1}{2}$.

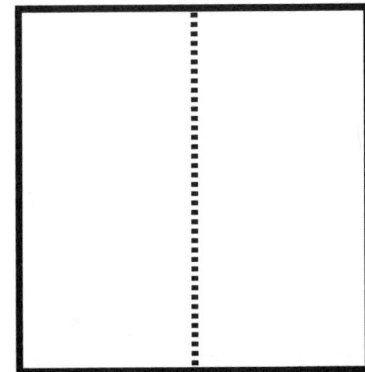

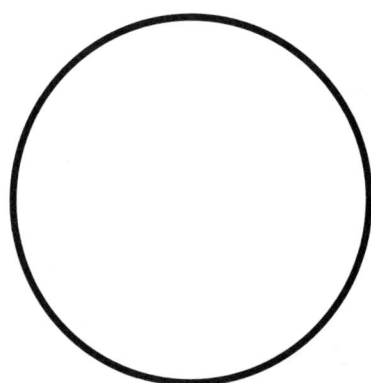

Show $\frac{1}{3}$.

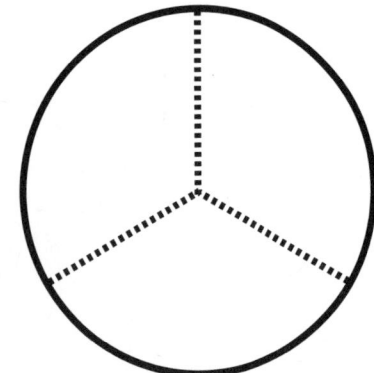

Duplicate fractions with Fraction Builder Strips. ◖ ❯

Name _____ Date _____

Identifying One Half and One Third

Ring the shapes that have equal parts.

Color these shapes to show $\frac{1}{2}$.

1.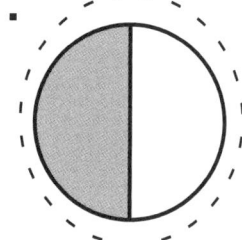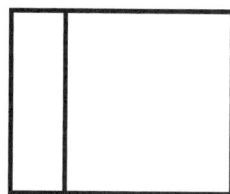

2.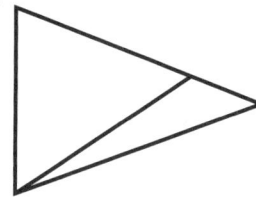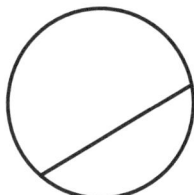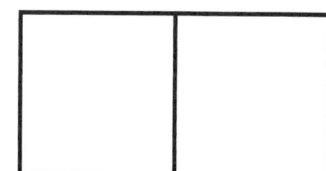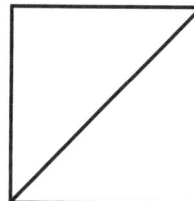

Ring the shapes that have equal parts.

Color these shapes to show $\frac{1}{3}$.

3.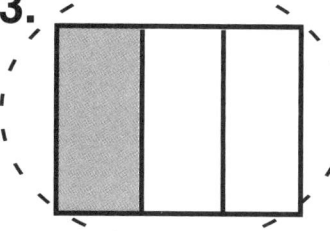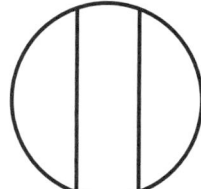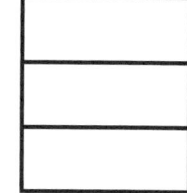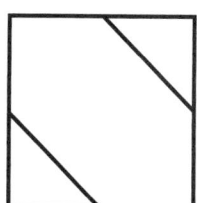

4.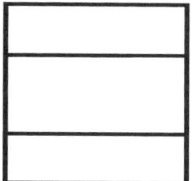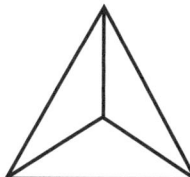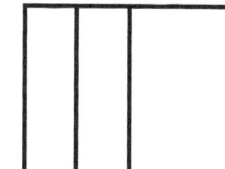

Demonstrate fractions with Overhead Fraction Circles.

Name _____ Date _____

Identifying One Half and One Third

2 equal parts

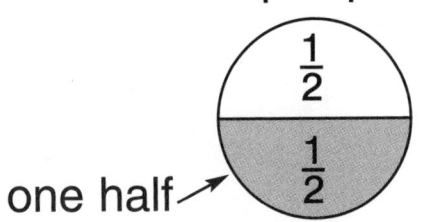

one half →

3 equal parts

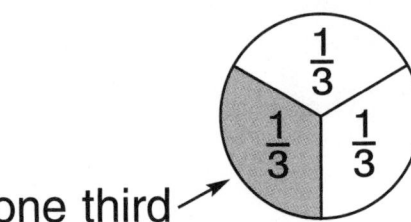

one third →

....

Color each shape that shows 2 equal parts.

1.

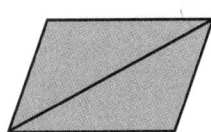 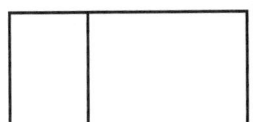  

2.

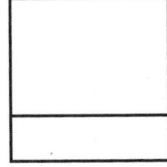 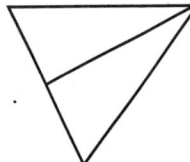 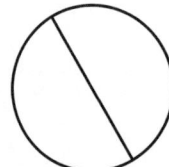 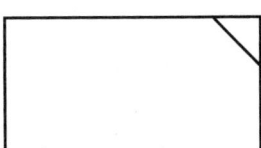

....

Ring each shape that shows 3 equal parts.

3.

 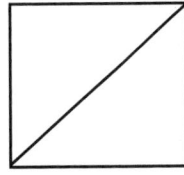 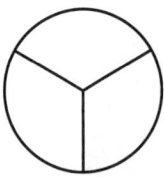 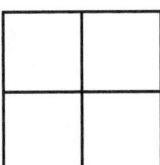 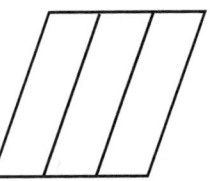

Duplicate fractions using Fraction Tiles. ⬜ ▶

Name _____ Date _____

Identifying One Fourth

Ring the fraction that each shape shows.

1.
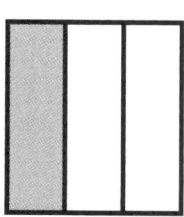

$\frac{1}{2}$ $\frac{1}{3}$ $\frac{1}{4}$

2.
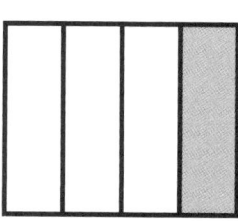

$\frac{1}{2}$ $\frac{1}{3}$ $\frac{1}{4}$

3.
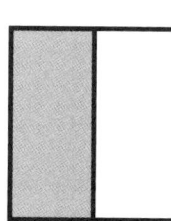

$\frac{1}{2}$ $\frac{1}{3}$ $\frac{1}{4}$

4.
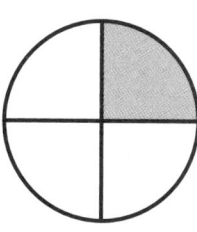

$\frac{1}{2}$ $\frac{1}{3}$ $\frac{1}{4}$

5.
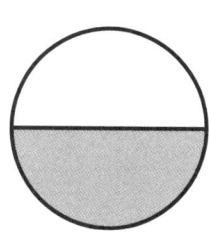

$\frac{1}{2}$ $\frac{1}{3}$ $\frac{1}{4}$

6.
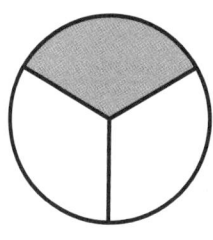

$\frac{1}{2}$ $\frac{1}{3}$ $\frac{1}{4}$

Find the shapes that show four equal parts. Color $\frac{1}{4}$.

7.

8. Which is greater, $\frac{1}{3}$ or $\frac{1}{4}$?

Ring the greater fraction.

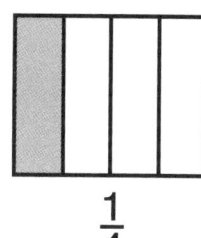
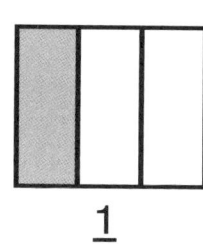

$\frac{1}{4}$ $\frac{1}{3}$

Duplicate fractions using Fraction Tiles.

Name _____   Date _____

Identifying One Fourth and One Fifth

4 equal parts

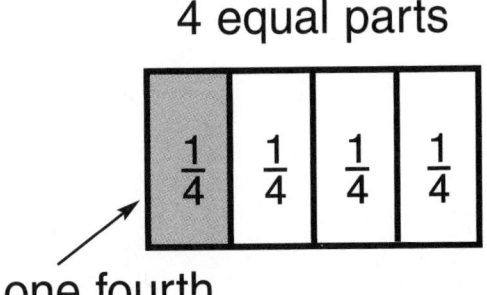

one fourth

5 equal parts

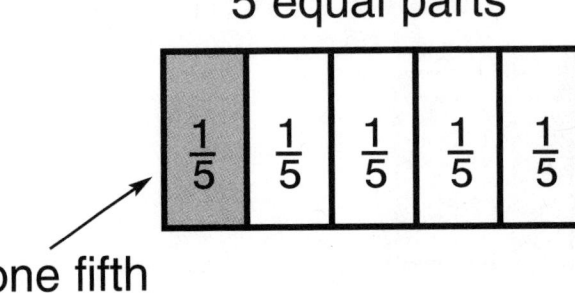

one fifth

Divide the shapes to show the fraction.

Show about $\frac{1}{4}$.

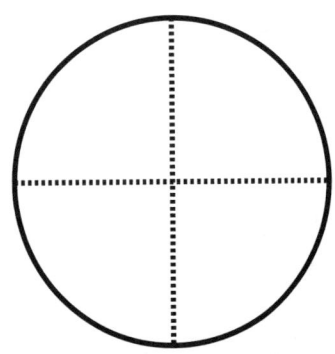

Show about $\frac{1}{5}$.

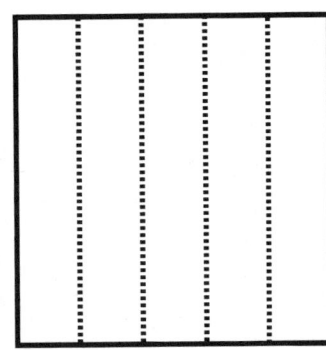

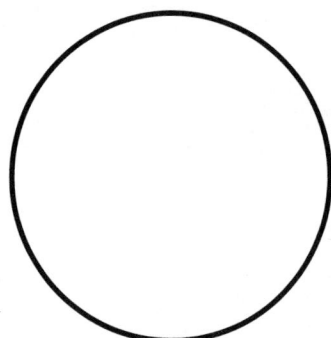

Duplicate fractions using Fraction Tiles.

Name _____ Date _____

Identifying One Fourth and One Fifth

Ring the shapes that have equal parts.

Color these shapes to show $\frac{1}{4}$.

1.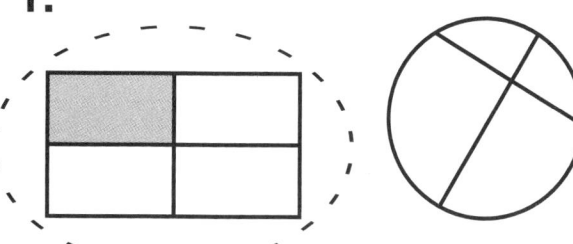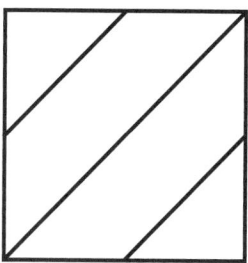

Ring the shapes that have equal parts.

Color these shapes to show $\frac{1}{5}$.

2.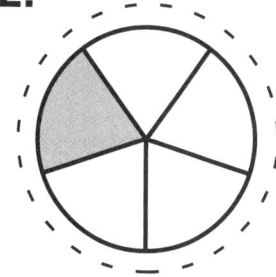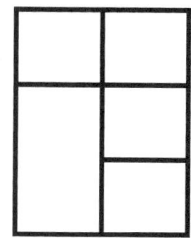

3. Write the fraction.

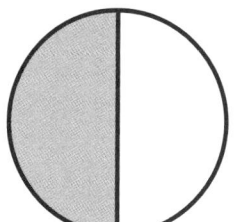 _____  _____ 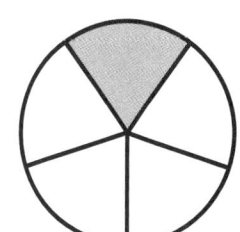 _____

Duplicate fractions using Fraction Tiles.

Name _____ Date _____

Identifying One Fourth and One Fifth

4 equal parts

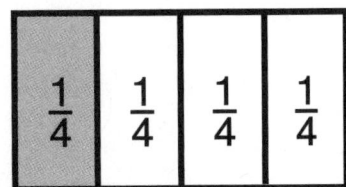

5 equal parts

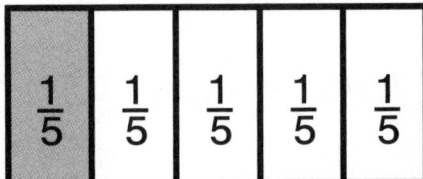

Ring the shapes that show $\frac{1}{4}$.

Put an X through shapes that show $\frac{1}{5}$.

1.

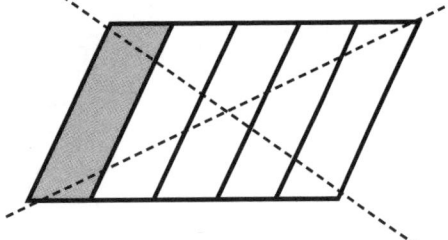

2.

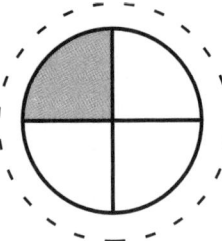

3.

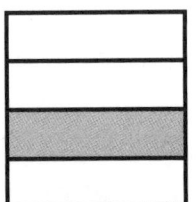

4.

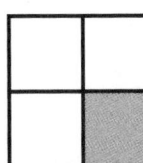

5.

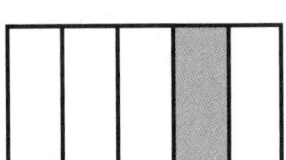

6.

7.

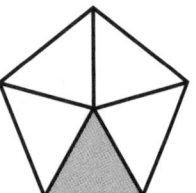

8.

Make and label fractions with Fraction Builder Strips. ◖❯

Name _____ Date _____

Reading Fractions

Ring the correct fraction.

1.
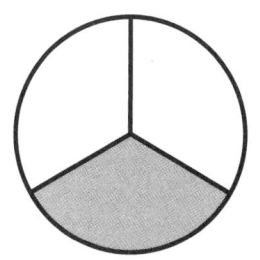
$\frac{1}{2}$ $\frac{1}{3}$

2.
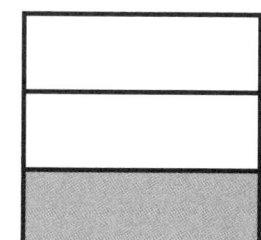
$\frac{1}{2}$ $\frac{1}{3}$

3.

$\frac{1}{2}$ $\frac{1}{3}$

4.
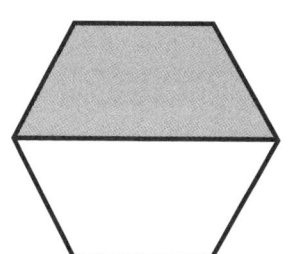
$\frac{1}{2}$ $\frac{1}{3}$

5.
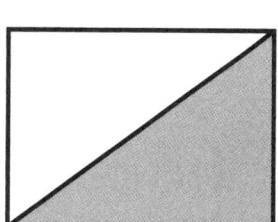
$\frac{1}{2}$ $\frac{1}{3}$

6.
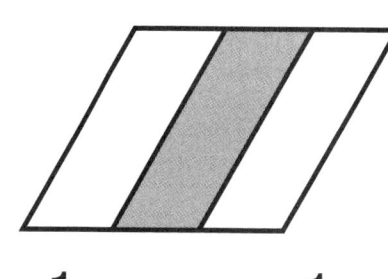
$\frac{1}{2}$ $\frac{1}{3}$

7.
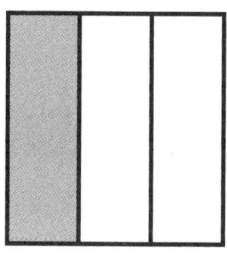
$\frac{1}{2}$ $\frac{1}{3}$

8.
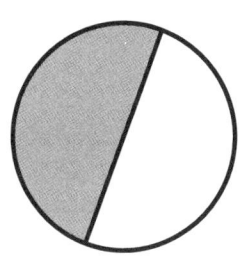
$\frac{1}{2}$ $\frac{1}{3}$

9.
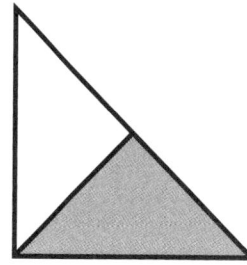
$\frac{1}{2}$ $\frac{1}{3}$

Make and label fractions with Fraction Builder Strips.

Name _____ Date _____

Identifying Fractions

Draw a line to the shape that shows the fraction.

1.

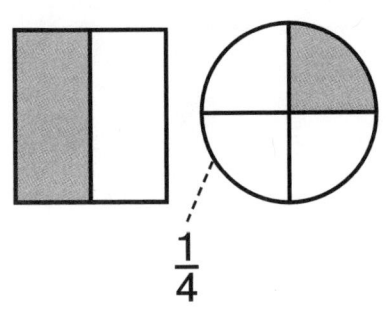

$\dfrac{1}{4}$

2.

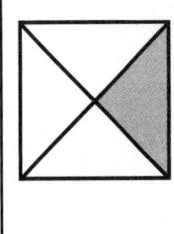

$\dfrac{1}{5}$

3.

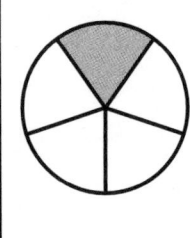

$\dfrac{1}{5}$

4.

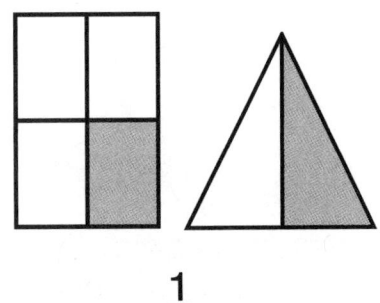

$\dfrac{1}{4}$

5.

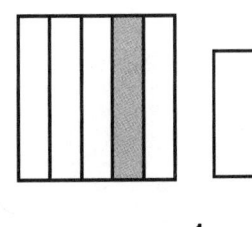

$\dfrac{1}{5}$

6.

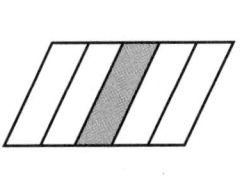

$\dfrac{1}{4}$

7.

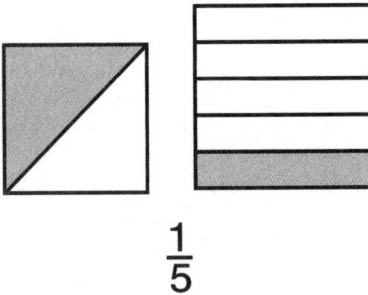

$\dfrac{1}{5}$

8.

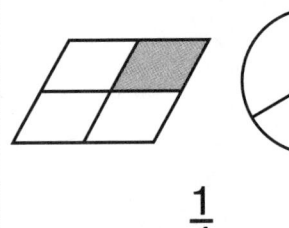

$\dfrac{1}{4}$

9.

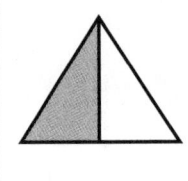

$\dfrac{1}{4}$

Check answers using Overhead Fractions.

Identifying Fractions

What part is shaded?
Ring the correct fraction.

1.

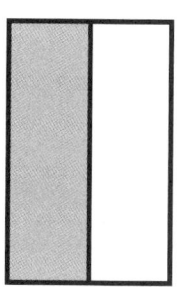

$\frac{1}{2}$ $\frac{1}{3}$ $\frac{1}{4}$

2.

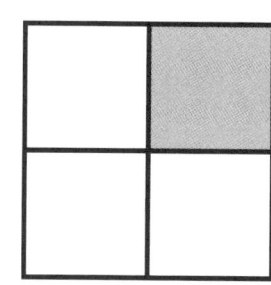

$\frac{1}{2}$ $\frac{1}{3}$ $\frac{1}{4}$

3.

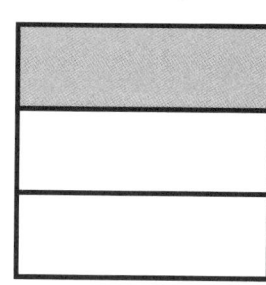

$\frac{1}{2}$ $\frac{1}{3}$ $\frac{1}{4}$

4.

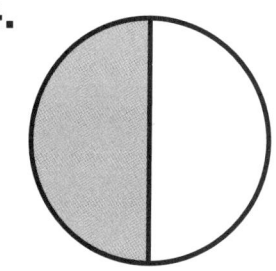

$\frac{1}{2}$ $\frac{1}{3}$ $\frac{1}{4}$

5.

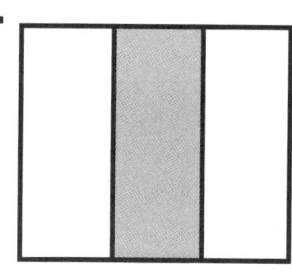

$\frac{1}{2}$ $\frac{1}{3}$ $\frac{1}{4}$

6.

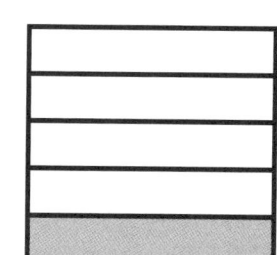

$\frac{1}{5}$ $\frac{1}{3}$ $\frac{1}{4}$

7.

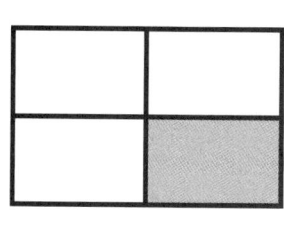

$\frac{1}{2}$ $\frac{1}{3}$ $\frac{1}{4}$

8.

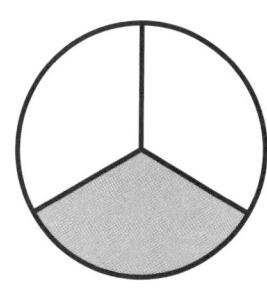

$\frac{1}{5}$ $\frac{1}{3}$ $\frac{1}{4}$

9.

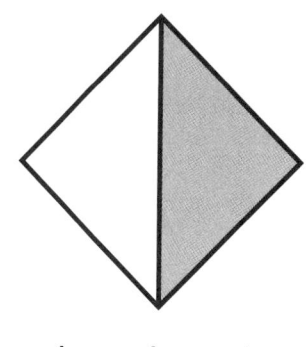

$\frac{1}{2}$ $\frac{1}{3}$ $\frac{1}{4}$

Make and label fractions with Fraction Builder Strips.

Name _____   Date _____

Fractions in Groups

Color to show each fraction.

1.

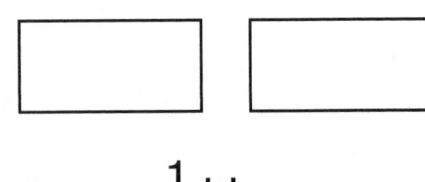

$\frac{1}{2}$ blue

2.

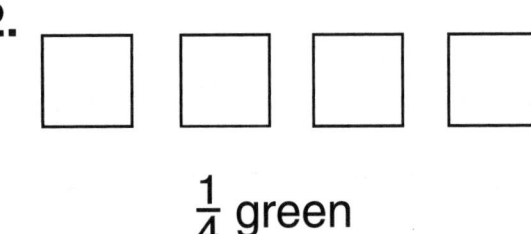

$\frac{1}{4}$ green

3.

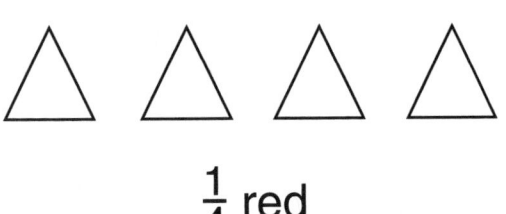

$\frac{1}{4}$ red

4.

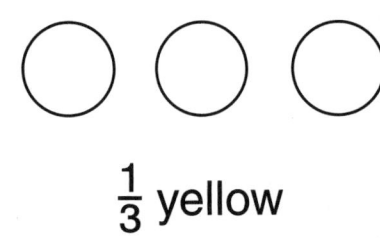

$\frac{1}{3}$ yellow

5.

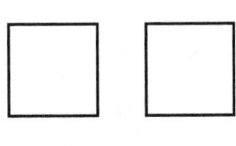

$\frac{1}{2}$ purple, $\frac{1}{2}$ orange

6.

$\frac{1}{3}$ green, $\frac{1}{3}$ red, $\frac{1}{3}$ blue

7. There were 3 bears in all.
Write the fraction that tells
what part of the group is left.

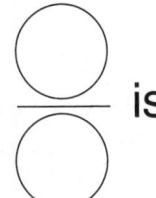 is left.

Check answers by duplicating problems with Attribute Blocks. ❓◻▶

Problem Solving

Think about sharing a pizza.

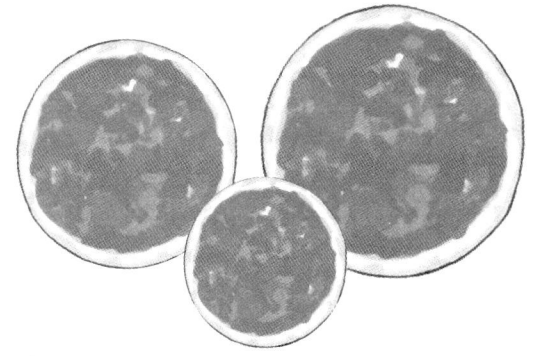

....

1. You want to give 3 children equal parts.
How could you cut the pizza? Ring it.

 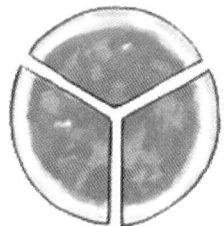 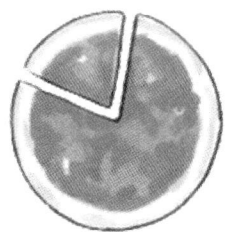 

---

2. You want to give 4 children equal parts.
How could you cut the pizza? Ring it.

  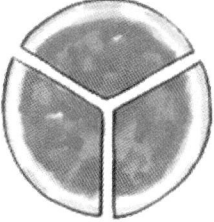 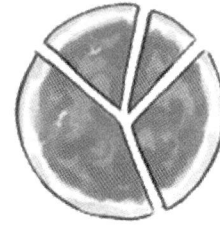

---

3. You want to give 2 children equal parts.
How could you cut the pizza? Ring it.

 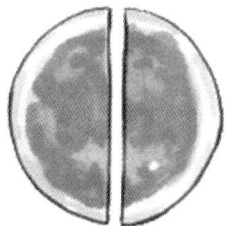 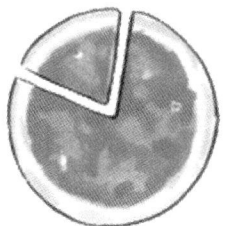 

Duplicate problems using Pizza Cardboards. ❓ ⬜ ▶

FRACTIONS USING PICTORIAL MODELS

Problem Solving

What is missing from each circle on the right?
Find the missing part. Draw lines to match.

1. $\frac{1}{2}$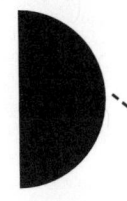

a.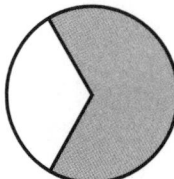

2. $\frac{1}{3}$

b.

3. $\frac{1}{4}$

c.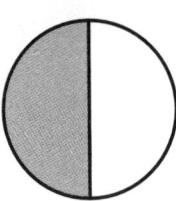

. . . .

How does the rest of each picture look?
Draw the missing part to complete.

4.

$\frac{1}{2}$ a triangle

5.

$\frac{1}{3}$ a pie

. . . .

Number Sense

6. There were 3 equal parts.
Ring the fraction that tells
what part of the pizza has
been eaten.

$\frac{1}{2}$ $\frac{2}{3}$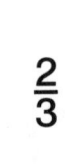

Demonstrate fractions with Overhead Fraction Circles. ❓⬜▶

Geometry for Primary Grade 1

Unit 1: Basic Ideas of Geometry

Assessment, P. 9
1. Color first and third figures; Cross out second and fourth figures, 2. Trace each side blue; Circle each corner; 3 sides; 3 corners, 3. Trace each side blue; Circle each corner; 8 sides; 8 corners, 4. triangle, 5. circle

P. 10
1. Color first and fourth figures; Ring second and third figures, 2. Color second and third figures; Ring first and fourth figures, 3. Color first and third figures

P. 11
1. Students will draw closed figures on Closed Tree and open figures on Open Tree, 2. Students will color inside closed shapes.

P. 12
Students will trace each side blue and circle each corner.

P. 13
1. Trace 4 sides blue; circle 4 corners; 4 sides; 4 corners, 2. Trace 3 sides blue; circle 3 corners; 3 sides; 3 corners, 3. Trace 4 sides blue; circle 4 corners; 4 sides; 4 corners, 4. Trace 5 sides blue; circle 5 corners; 5 sides; 5 corners, 5. Trace 3 sides blue; circle 3 corners; 3 sides; 3 corners, 6. Trace 6 sides blue; circle 6 corners; 6 sides; 6 corners

P. 14
1. Trace 3 sides blue; circle 3 corners; 3 sides; 3 corners, 2. Trace 4 sides blue; circle 4 corners; 4 sides; 4 corners, 3. Trace 8 sides blue; circle 8 corners; 8 sides; 8 corners, 4. Trace 5 sides; circle 5 corners; 5 sides; 5 corners, 5. Trace 3 sides blue; circle 3 corners; 3 sides; 3 corners, 6. Trace 4 sides blue; circle 4 corners; 4 sides; 4 corners

P. 15
1. 4 sides; 4 corners; 4 square corners, 2. 3 sides; 3 corners; 0 square corners, 3. 4 sides; 4 corners; 4 square corners, 4. 4 sides; 4 corners; 2 square corners, 5. 3 sides; 3 corners; 1 square corner, 6. 5 sides; 5 corners; 2 square corners

P. 16
1. square, 2. rectangle, 3. triangle, 4. circle

Unit 2: Plane Figures

Assessment, P. 17
1. Ring circle, 2. Ring triangle, 3. Ring rectangle, 4. Ring square, 5. Ring third circle

P. 18
1. square, 2. circle, 3. rectangle, 4. triangle

P. 19
Students will accurately color the shapes.

P. 20
Students will match the shapes.

P. 21
1. Color last rectangle, 2. Color cube, 3. Color last circle, 4. Color triangle

P. 22
1. Color third and sixth triangles, 2. Color first and fifth circles, 3. Color third and fifth rectangles, 4. Color second and sixth squares

P. 23
1. Ring second triangle, 2. Ring last rectangle, 3. Ring first square, 4. Ring second circle

P. 24
1. Ring first circle, 2. Ring first square, 3. Ring third triangle, 4. Ring rectangle, 5. Ring circle

P. 25
1. Ring first two shapes, 2. Ring first, fourth, fifth shapes, 3. Ring first, third, fifth shapes, 4. Ring first three shapes, 5. Smaller, Same, Bigger

P. 26
Students will correctly draw shapes.

P. 27
Students should draw identical figures.

P. 28
Students will correctly draw shapes.

P. 29
Students will fill in bar graph to show 8 triangles, 4 squares, 7 circles, 5 rectangles

P. 30
1. rectangle, 2. triangle, 3. circle

P. 31
1. Ring second box, 2. Ring first box, 3. Ring first box, 4. Ring second box

P. 32
1.

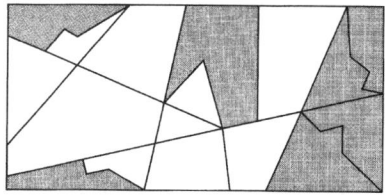

2.

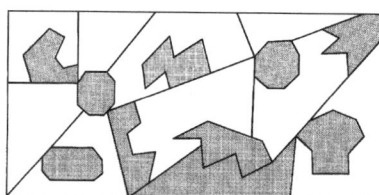

P. 33
Students will accurately draw shapes.

Unit 3: Solid Figures

Assessment, P. 34
1. b, 2. Match cylinder to soup pot, 3. Ring first shape, 4. Ring third shape

P. 35
1. 1. d, 2. c, 3. a, 4. b, 5. 8

P. 36
1. d, 2. c, 3. b, 4. a

P. 37
1. Ring box, 2. Ring ball, 3. Ring construction cone, 4. Ring carrot can

P. 38
Students will color figures as directed.

Cone	4	Cylinder	3
Sphere	5	Cube	3

P. 39
1. Color gift and building block red, 2. Color orange and ball blue, 3. Ring glass and trash can, 4. X funnel and cone by sink, 5. Underline toy box and wood box

P. 40
1. Match ball, sphere, globe
2. Match block, cube, ice cube
3. Match jar, cylinder, glass
4. Match megaphone, cone, ice cream cone

P. 41
Color the megaphone and hat blue, Color the TV and block red, Color the paint can and soup can yellow, Color the ball and orange green

P. 42
1. Students will accurately color the picture, 2. Match the sphere to the ball, 3. Match the cylinder to the pot, 4. Match the cone to the cone, 5. Match the cube to the box

P. 43
Students will fill in bar graph to show 8 cubes, 3 cones, 5 cylinders, 6 boxes

P. 44
1. Ring first, second, third, fourth shapes, 2. X first, third, fifth shapes, 3. Color first, second, third, fourth shapes, 4. cone

P. 45
1. Color first, second, fourth, fifth shapes, 2. Color second, third, fourth shapes, 3. Color first, second, fourth, fifth shapes, 4. Ring first, second, third shapes

Unit 4: Symmetry

Assessment, P. 46
1. yes, 2. no, 3. yes, 4. no, 5. yes, 6. yes

P. 47
1. no, 2. yes, 3. yes, 4. no, 5. yes, 6. yes

P. 48
Students will draw lines of symmetry.

P. 49
1-6. Students will draw lines of symmetry, 7. Ring the cup

P. 50
1. no, 2. yes, 3. yes, 4. no, 5-7 Students will draw lines of symmetry

P. 51
1. no, 2. yes, 3. yes, 4. no, 5. yes, 6. no, 7-9 Students will draw lines of symmetry

ANSWER KEY

Geometry for Primary Grade 1

Unit 5: Measurement

Assessment, P. 52
1. about 7 cm, 2. about 5 inches,
2. 4 centimeters; 4 centimeters;
3 centimeters; 11 centimeters in all,
4. 5 square centimeters

P. 53
1. 1 cm, 2. 3 cm, 3. 6 cm

P. 54
1. about 12 cm, 2. about 6 cm, 3. Ring
the first pencil, X the second pencil

P. 55
1. 4 cm, 2. 7 cm, 3. 10 cm, 4. 6 cm,
5. 8 cm

P. 56
1. about 13 cm, 2. about 12 cm,
3. about 4 cm, 4. about 7 cm

P. 57
1. 3 cm, 2. 2 cm, 3. about 12 cm,
4. about 6 cm, 5. about 8 cm

P. 58
1. 1 inch, 2. 4 inches, 3. 6 inches

P. 59
1. about 3 inches, 2. about 4 inches,
3. about 5 inches, 4. about 6 inches

P. 60
1. 3 inches long, 2. 5 inches long, 3. about
4 inches long, 4. about 6 inches long

P. 61
1. 4 inches long, 2. 5 inches long,
3. 2 inches long, 4. 6 inches long

P. 62
Drawings should be as accurate as
possible.

P. 63
1. 2 centimeters; 2 centimeters;
3 centimeters; 7 centimeters in all,
2. 4 centimeters; 4 centimeters;
3 centimeters; 11 centimeters in all

P. 64
1. 9 + 10 + 5 = 24 centimeters
2. 5 + 5 + 5 + 5 = 20 centimeters

P. 65
1. 2 square centimeters, 2. 4 square
centimeters, 3. 10 square centimeters,
4. 8 square centimeters, 5. 3 square
centimeters, 6. 4 square centimeters

P. 66
1. 10 centimeters, 2. 14 centimeters,
3. 12 square centimeters, 4. 10 square
centimeters, 5. 5 square centimeters,
6. 9 square centimeters

P. 67
Answers will vary. Make sure students use
the same number of square centimeters.

Unit 6: Patterns

Assessment, P. 68
1. circle, 2. diamond, 3. diamond,
4. rectangle

P. 69
1. square, 2. circle, 3. rectangle

P. 70
1. square, 2. diamond, 3. diamond,
4. square

P. 71
1. 2 square, 2. 2 circle, 3. 3 square,
4. 8 circle

P. 72
1. b, 2. a, 3. c, 4. b

P. 73
1. square, 2. 4-sided polygon, 3. 4-sided
polygon, 4. triangle, 5. triangle

Unit 7: Fractions Using Pictorial Models

Assessment, P. 74
1. Ring first and third figures,
2. Ring first and second figures,
3. Ring first and third figures,
4. $\frac{1}{2}$, 5. $\frac{1}{5}$, 6. $\frac{1}{4}$

P. 75
1. Color first triangle, 2. Color first
rectangle, 3. Color first square, 4. Color
first circle, 5. Trace and ring first hexagon,
6. Trace and ring first triangle

P. 76
1. Ring second figure, 2. Ring second
figure, 3. Ring second figure, 4. Ring
first figure, 5. Ring first figure, 6. Ring
second figure

P. 77
1. yes, 2. yes, 3. no, 4. yes, 5. no, 6. yes,
7. Color second figure, 8. Color second
figure, 9. Color first figure,
10. Color first figure

P. 78
1. Trace and color first figure, 2. Trace and
color first figure, 3. Trace and color second
figure, 4. Trace and color second figure, 5.
Trace and color first figure,
6. Trace and color first figure

P. 79
Students should accurately show divide
lines.

P. 80
1-4. Students will accurately divide and
color figures, 5. Color second and third
figures

P. 81
1. $\frac{1}{3}$, 2. $\frac{1}{2}$, 3. $\frac{1}{3}$, 4. $\frac{1}{2}$, 5. $\frac{1}{3}$, 6. $\frac{1}{3}$, 7. Color $\frac{1}{3}$
of second and third figures, 8. Color $\frac{1}{3}$ of
third figure, 9. $\frac{1}{2}$

P. 82
Students will draw lines to show about
$\frac{1}{2}$ and $\frac{1}{3}$.

P. 83
1. Ring and color first and second figures,
2. Ring and color third and fourth figures,
3. Ring and color first and third figures,
4. Ring and color second and fourth figures

P. 84
1. Color first and third figures, 2. Color
second and third figures, 3. Ring first,
third, and fifth figures

P. 85
1. $\frac{1}{3}$, 2. $\frac{1}{4}$, 3. $\frac{1}{2}$, 4. $\frac{1}{4}$, 5. $\frac{1}{2}$, 6. $\frac{1}{3}$, 7. Color $\frac{1}{4}$ of
second and third figures, 8. $\frac{1}{3}$

P. 86
Students will draw lines to show about $\frac{1}{4}$
and $\frac{1}{5}$.

P. 87
1. Ring and color first and third figures,
2. Ring and color first, second, and third
figures, 3. $\frac{1}{2}$, $\frac{1}{4}$, $\frac{1}{5}$

P. 88
1. X, 2. Ring, 3. Ring, 4. Ring, 5. X, 6. X, 7.
X, 8. Ring

P. 89
1. $\frac{1}{3}$, 2. $\frac{1}{3}$, 3. $\frac{1}{2}$, 4. $\frac{1}{2}$, 5. $\frac{1}{2}$, 6. $\frac{1}{3}$, 7. $\frac{1}{3}$, 8. $\frac{1}{2}$,
9. $\frac{1}{2}$

P. 90
1. Second figure, 2. Second figure,
3. First figure, 4. First figure, 5. First figure,
6. Second figure, 7. Second figure,
8. First figure, 9. Second figure

P. 91
1. $\frac{1}{2}$, 2. $\frac{1}{4}$, 3. $\frac{1}{3}$, 4. $\frac{1}{2}$, 5. $\frac{1}{3}$, 6. $\frac{1}{5}$, 7. $\frac{1}{4}$, 8. $\frac{1}{3}$, 9.
$\frac{1}{2}$

P. 92
1-6. Students accurately color fractions,
7. $\frac{1}{3}$

P. 93
1. Ring second pizza, 2. Ring first pizza,
3. Ring second pizza

P. 94
1. c, 2. a, 3. b, 4. Students complete
picture, 5. Students complete picture, 6. $\frac{2}{3}$